T0231669

ENERGY INTERMITTENCY

ENERGY INTERMITTENCY

BENT SØRENSEN
Roskilde University
Roskilde, Denmark

CRC Press
Taylor & Francis Group
Boca Raton London New York

CRC Press is an imprint of the
Taylor & Francis Group, an **informa** business

CRC Press
Taylor & Francis Group
6000 Broken Sound Parkway NW, Suite 300
Boca Raton, FL 33487-2742

First issued in paperback 2019

© 2015 by Taylor & Francis Group, LLC
CRC Press is an imprint of Taylor & Francis Group, an Informa business

ISBN-13: 978-1-4665-1606-9 (hbk)
ISBN-13: 978-1-138-37469-0 (pbk)

Library of Congress Cataloging-in-Publication Data

Sorensen, Bent, 1941-
    Energy intermittency / author, Bent Sorensen.
        pages cm
    Includes bibliographical references and index.
    ISBN 978-1-4665-1606-9 (hardback)
    1. Electric power system stability. 2. Energy storage. 3. Electric power failures--Prevention. I. Title.

TK1010.S67 2014
621.31'2--dc23   2014028234

Visit the Taylor & Francis Web site at
http://www.taylorandfrancis.com

and the CRC Press Web site at
http://www.crcpress.com

Other Books by the Author

A History of Energy. Northern Europe from the Stone Age to the Present Day, 2011/2012.

Hydrogen and Fuel Cells, 2nd ed., 2011/2012 (1st ed., 2005).

Life-Cycle Analysis of Energy Systems: From Methodology to Applications, 2011.

Renewable Energy Reference Book Set (ed., 4 volumes of reprints), 2010.

Renewable Energy: Physics, Engineering, Environmental Impacts, Economics and Planning, 4th ed., 2010 (Previous editions 1979, 2000, and 2004).

Renewable Energy Focus Handbook (with Breeze, Storvick, Yang, Rosa, Gupta, Doble, Maegaard, Pistoia, and Kalogirou), 2009.

Renewable Energy Conversion, Transmission and Storage, 2007.

Life-Cycle Analysis of Energy Systems (with Kuemmel and Nielsen), 1997.

Blegdamsvej 17, 1989.

Superstrenge, 1987.

Fred og frihed, 1985.

Fundamentals of Energy Storage (with Jensen), 1984.

Energi for fremtiden (with Hvelplund, Illum, Jensen, Meyer and Nørgård), 1983.

Energikriser og Udviklingsperspektiver (with Danielsen), 1983.

Skitse til alternativ energiplan for Danmark (with Blegaa, Hvelplund, Jensen, Josephsen, Linderoth, Meyer and Balling), 1976.

More information on the author's work is available at http://energy.ruc.dk

Contents

Preface

Energy variation and intermittency are key issues for all existing energy systems and particularly for new energy systems such as those based on renewable energy sources that flow irregularly. Variations in source flow can hinder the ability to match supply and demand, even if the source drop is not to zero, and so can insufficient transmission capacity or insufficient energy conversion equipment. This is why the word *intermittency* is used to cover both situations, i.e., those with no flow into the system and those situations where variations in conjunction with the system structure make the supply fall short of the desired demand, thus causing intermittency on the demand side unless countermeasures are undertaken.

Despite its importance, intermittency has received only sporadic mention in the energy literature. This book sets out to remedy that situation by describing the causes of intermittency as well as potential countermeasures across the board for conventional energy systems based on fossil or nuclear fuels as well as for renewable energy systems. Three types of solutions are discussed: trade arrangements, such as by power grid interconnections; active energy storage at different scales; and demand management. After a general technical description of the options, a number of case studies are presented to show how to solve the problems and furnish a resilient, working energy system for all types of energy resources, including systems with 100% reliance on intermittent energy sources. The case studies are for regions in North America, Europe, and Southeast Asia, where the combination of population size and resource availability in countries such as Japan, Korea, and China puts the construction of a sustainable energy system to a severe test.

The book is written at a broadly accessible level and should cater to energy planners in government and industry, to technical people involved in energy science, and to the broad range of political and grassroots communities engaging in discussions of the energy issues facing all regions of the world over the coming decades.

Bent Sørensen

Gilleleje, February 2014

boson@ruc.dk

About the Author

Bent Sørensen is professor emeritus at Roskilde University. He previously held academic positions at the Universities of New South Wales (Australia), Grenoble (France), Kyoto (Japan), Copenhagen (Denmark), Yale (Connecticut), and Berkeley (California), and at the National Renewable Energy Laboratory (Colorado), and he has been technical director of Denmark's largest engineering firm. He served as advisor for OECD (Organisation for Economic Co-operation and Development) and several governments and UN agencies. In addition, he was a lead author for the IPCC (Intergovernmental Panel on Climate Change) Second Assessment Report. He is the recipient of several awards and honours, including the European Solar Prize.

1

INTRODUCTION

Current energy systems are to a large extent based on fossil and nuclear fuels, and thus they are subject to rising long-term concerns over the associated emissions (greenhouse gases for combustion of fossil fuels and the risk of accidental releases of radioactive substances for nuclear reactors) and, in the medium term, issues of resource depletion. Alternatives without such problems include the renewable energy sources derived from the disposition of solar energy on the Earth. Several renewable energy sources are characterized by an intermittent flow, such as that of wind energy, which depends on passing weather systems, or that of solar radiation, which is absent at night and variable on cloudy days, in addition to the seasonal variation at higher latitudes. As regards energy systems, intermittency and variability should be seen in the light of energy production matching energy demand. The title of this book only mentions intermittency, but the book intends to deal with both intermittency and variability, which in many cases leads to intermittency if the variability causes inability to cover the demand at a given moment in time. The system may therefore be unable to follow changes in demand, even if the energy source flow is not intermittent. There are several ways of handling the issues posed by supply-demand mismatch, and these are the issues this book examines in detail.

For fuel-based energy production, such as by fossil fuels or fuels of fresh biomass origin as well as by nuclear fuels, a storage-before-conversion option immediately offers itself. However, even here, there may be intermittency issues, because retrieving and making use of stored fuels may involve a time delay, for example, caused by the start-up times for facilities such as coal or nuclear power plants, which can be a problem in cases of unexpected demand change. Generally, many such problems may be addressed by planned production, for example, by operating several power plants at less than full capacity, so that regulation up and down is possible. For many types of nuclear power plants, such regulation is undesirable because it increases the risk of instability, and these plants are therefore often operated at a fixed output level. Handling intermittency may involve any—or a combination—of energy storage, production planning, and demand planning.

Production planning may consist of operating a system with surplus capacity, as in the example of fossil fuel power plants, but it also extends to a collaboration between more than one "system," usually implying collaboration agreements or import–export between different energy utilities, enabling them to borrow from each other in cases of deficit and to get rid of surpluses. The energy form used for such trade agreements may be electricity or piped heat or fuel (gas or oil pipelines), or the storable fuels may of course be traded by use of ships, trains, or other vehicles. A scheme employing both energy storage and power exchange between operators is the collaboration between a hydropower system based on large upper storage reservoirs and a wind or photovoltaic power system, which may import hydropower at times of insufficient wind or solar production and then pay it back in periods of surplus production. The only extra cost is to have the hydro turbines rated at so high a power level that they can furnish the additional power for export, which is usually a very small part of the cost of constructing a reservoir-based hydro system. (The key cost components are usually dams and environmentally acceptable management of the flooded reservoir areas.) When there is a wind or solar surplus, they satisfy demand, and the hydro turbines are regulated down correspondingly. This solution is clearly both less expensive and more energy efficient than the pure storage option, for example, where surplus wind power is used to pump water upwards for later use in hydropower generation. If the reservoirs deliver seasonal storage (as they do in areas where reservoirs are filled in spring by melting snow and emptied over the rest of the year, e.g., in Norway or Canada), then the disturbance of the reservoir water level by serving as backup for, say, a wind energy system is often only a few centimeters change in average filling level.

Demand management consists of performing energy-intensive jobs at specific times that suit the capability of the energy system. This could involve having a washing machine stand filled and awaiting the time when a surplus of power is available on the electricity grid. The control could be by a combination of signals sent by the power company over the grid and decisions made by a computer program in the home. There are other management options furnished by the many battery-operated portable devices currently in use. Recharging batteries can be done when it suits the electricity supply system, creating a store of charged batteries to be used over a following period of time. A combination of demand management and power exchange is offered by the diurnal variations in demand characteristic in most user communities. For example, the peak use of electricity may happen around 5 p.m. in a number of locations situated in different time zones, allowing peak consumption to be covered from neighboring longitude zones that are off-peak at the time in question. This was exploited quite early in the former Soviet Union, presumably due to the presence of only one electric utility company across the several time zones between Europe and Vladivostok. However, this idea can also be put to use in other places if the relevant utility companies can agree on a suitable power-exchange arrangement.

When simple import–export arrangements and the always limited demand-management options (some tasks cannot be postponed) are not sufficient, dedicated use of energy stores must be explored, even if storage options are often more expensive than management solutions. The case of batteries for portable equipment (smartphones, portable computers, drilling machines, lawn mowers, and so on) already shows examples of technology that consumers find attractive and are willing to pay the extra price of battery operation to get. A straight extrapolation of these ideas is to create smart buildings, where energy production (rooftop solar cells) and energy import (district heating, electric power) are combined with storage at the building level, such as basement or underground hot-water and hydrogen stores operated in conjunction with reversible fuel cell units. The next step would be to take advantage of the cost reductions often associated with communal energy stores, as compared with single-building integrated ones. This can be a complex issue, because the space taken up by the communal installation, although smaller than that of the individual solution, may be seen as less easy to accommodate into city planning. Most such installations are today placed on marginal land at the outskirts of cities, e.g., in conjunction with district heating systems.

2

INTERMITTENCY DEPENDENCE ON TYPE OF ENERGY SYSTEM

Energy systems range from small, individual ones, such as a three-stone cooking fire in a remote African village, to over local systems not communicating with their environments (for example, the energy system of an isolated city without grid or pipeline connection to other energy systems), to large interconnected energy systems with many built-in options for power exchange and other trade of fuels or processed energy. These systems also differ in their response to intermittency of primary energy inputs.

Consider first an individual energy system, as depicted in an idealized form in Figure 2.2 (the symbols used in this and the two following illustrations are explained in Figure 2.1). The primary energy sources may be purchased fuels, purchased energy stores (such as batteries), or the energy may come from renewable energy conversion equipment (wind turbines, solar thermal or photovoltaic panels, draft animals). The demands would typically be in the form of electric power for consumer devices, energy for space heating or cooling, or use in equipment for cooking, for food preservation, for operating tools in local manufacture, and for mobility. In-house or extremely local distribution networks of power may be present, and in regions with a space-heating demand, possibly also a heat distribution system (such as water pipes or air ducts), but rarely other types of energy transfer.

Figure 2.3 shows schematically an example of a typical conventional energy system for a supply area that may be a utility servicing area such as a city or a region. The example has energy input from fuels (e.g., fossil or nuclear) converted into grid electricity, district heat (e.g., through combined power and heat plants), and pipeline-quality gases such as town gas, natural gas, or hydrogen, and inputs from renewable sources (wind, solar, biomass), of which a part goes through energy stores for heat or for regeneration of high-quality

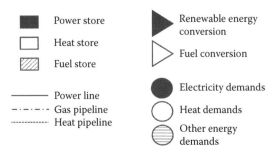

Figure 2.1 Symbols used in Figures 2.2–2.4. By "power store" is understood an energy store capable of regenerating the energy in the form of electricity.

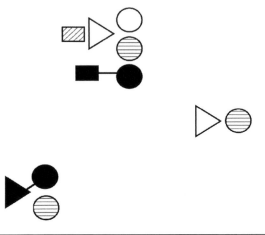

Figure 2.2 Example of independent consumers without a common grid, of which some use fuels for energy generation (say fuelwood for cooking), perhaps also for space heating (top), and where renewable systems (such as solar panels, used when solar radiation is available) or batteries (top) are used to generate electric power.

energy, mostly in the form of electricity. The system depicted thus has a network of transmission lines for electric power, gas, and heat. Local systems (e.g., in buildings) differ, and some have only power input from the transmission grid, while others are also connected to the district heating lines. Just five examples are shown: The top one on the left is characterized by distributing both electricity through power lines and hot water through a district heating grid, thereby covering power and heat loads, while other loads such as industrial process heat can be covered directly from the (natural) gas transmission lines.

The middle local area represented on top has its own fuel-based production of heat, based, e.g., on oil or wood fuel, while the local system to the right produces heat based on the gas grid. At the bottom, two local systems with their own energy production from renewable energy are illustrated. The one to the left produces only electric power (photovoltaic panels), covering part of the load, assuming the rest to be covered from the grid, while that on the right also has its own heat production (solar thermal panels or utilization of the waste heat from the photovoltaic conversion) and a heat store (say a rock store under the building) for covering the seasonal mismatch between solar radiation and loads of space heating and hot water.

A system such as the one illustrated in Figure 2.3 is unable to cope with serious problems such as loss of major central generating facilities (by accidents or other failure) or disruption of major transmission lines (by natural catastrophes or otherwise). Dealing with such important

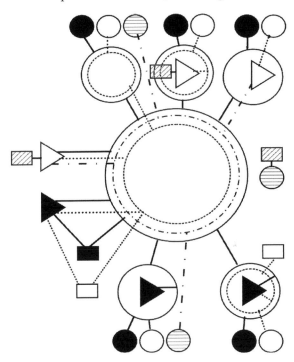

Figure 2.3   Example of a single-area (regional) energy supply system, employing general transmission networks for power, heat, and gas, as well as various combinations of local (e.g., in-building) energy distribution grids. The system employs both centralized and decentralized generation equipment based on (stored) fuels or on renewable energy sources with heat stores as well as stores capable of regenerating electric power. Detached from the grid is a fuel-based transportation sector, but some transportation demand may be fulfilled by use of gaseous (natural gas or hydrogen) fuels from the pipelines or electricity (for trains and other electric vehicles) from the grid. Local areas or buildings may have additional renewable energy production or energy conversion by boilers, heat pumps, or fuel cells.

but hopefully infrequent events may involve hooking the system up to neighboring energy supply systems, allowing transmission of power and other energy forms in case of problems with the components of the regional system. Only some problems can be dealt with in this way. For instance, cases of losing the integrity of major transmission lines can of course not be resolved by new import options if there is no way of getting the imported energy to the customers. On the other hand, connecting individual systems can offer advantages beyond the emergency uses. Once the expense of establishing interconnections between regional (or national) networks has been committed, it would be foolish not to make the fullest use of the connectivity. This means power exchange and other energy transfer that improves system stability, and it could mean greatly improved options for dealing with intermittent energy sources (OECD/IEA 2005).

Some of these options are exemplified in Figure 2.4, showing a schematic case of interconnected energy systems, of which some are fully relying on renewable energy resources and thus exposed to all the issues of intermittency associated with sources such as wind or solar radiation. The lower right system can use its fossil fuel components to counteract dips and peaks in matching the renewable energy components to demand, and quite easily manage the matching process, because all forms of energy transmission lines are available. In the lower left system, the centralized production variability is regulated by use of stores for both heat and high-quality energy, meaning stores like pumped hydro, batteries, compressed air, or hydrogen, all characterized by being able to retrieve the stored energy in the form of electricity. The two local systems illustrated above the central grids have local energy production by means of fuel cells, expected in the future to replace the natural gas cogeneration units used in some places today (Sørensen 1983, 1984). The system to the left (say for a building) has only one-way generation, converting hydrogen into heat and power for the building. The system to the right uses two-way (reversible) fuel cells, either taking electric power from the grid and converting it to hydrogen for filling into a vehicle, or for storing to later regeneration or usage for transportation, and with the associated "waste" heat used to fulfill needs in the building, or—in reverse mode—taking hydrogen from the local store and converting it to electricity when the primary production (from solar radiation or wind) through the grid is

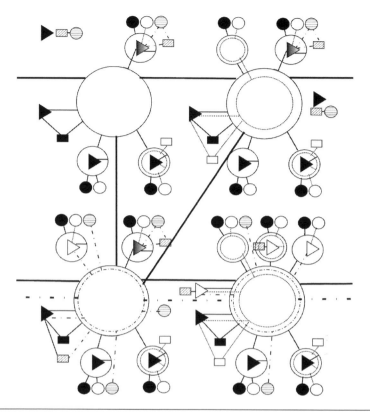

Figure 2.4   Example of interconnected energy systems with transmission grid and pipeline connections to additional international partners. The system at the bottom right is the one depicted in Figure 2.3, and the one at the bottom left is a similar system but without district heating lines, with all energy production coming from renewable energy. It uses fuel cells in local areas, either fed by hydrogen from central transmission pipelines (upper left local system) or producing its own hydrogen by reversible fuel cells (the triangle with graded filling of the upper right local system) with use of local hydrogen stores. The systems at the top have no gas pipeline transmission lines, but the one on the right has a district heating network. Both have unattached biofuels systems providing energy for vehicles in the transportation sector.

insufficient. In this direction of operation, the "waste" heat generated may also be made useful.

Dealing with intermittency in the way described for the upper right-hand local system is also found in the two general systems shown at the top of Figure 2.4, now with only electricity grid or with additional district heating, but no gas pipelines. The heat stores placed locally in these systems must be capable of seasonal storage in order to be useful for space heating at higher latitudes. The local hydrogen production is not considered sufficient for covering all transportation needs, and an independent production of biofuels is assumed to be available. Should further intermittency-handling capability for the entire system be required, the use of biofuels would in this case offer an additional option for storage and regeneration of high-quality energy.

Figure 2.5 illustrates a conventional fuel-based energy system, including typical indications of diurnal and seasonal time variations in supply and demand, thereby signaling a need to handle variations and intermittency even for a system based on storable fuels. The reasons

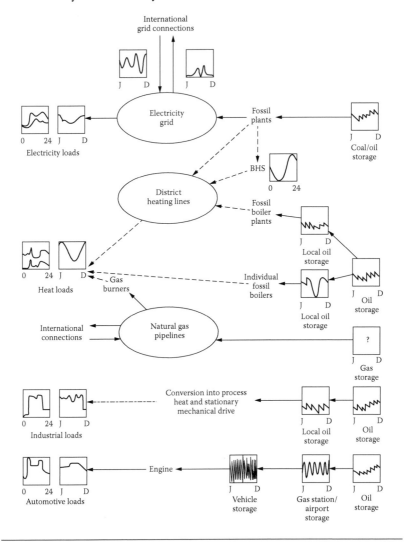

Figure 2.5 Intermittency in a conventional fuel-based energy system is illustrated by giving load, supply, and storage filling level variations on a diurnal scale (denoted 0 to 24, in hours) and on a seasonal scale (denoted J to D, monthly from January to December). (From Jensen and Sørensen [1984]. Used with permission.)

for this include the inertia of generation units, where natural-gas- and oil-fired units may change output at short-term notice, but where coal-fired power stations require longer times, ranging from minutes when already operating at partial load to hours if a cold start of a new unit is required. These problems are traditionally handled by grid interconnections or by use of peak-power units (usually gas turbines) for electricity and by buffer heat stores for heat. Figure 2.6 gives a similar overview of

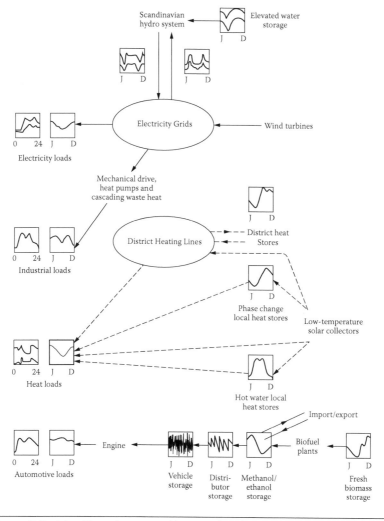

Figure 2.6 Intermittency in a renewable energy-based energy system (proposed for the Scandinavian regions) is illustrated by giving load, supply, and storage filling level variations on a diurnal scale (denoted 0 to 24, in hours) and on a seasonal scale (denoted J to D, monthly from January to December). (From Jensen and Sørensen [1984]. Used with permission.)

a renewable energy system. Apart from the biofuels component, which can be stored like fossil fuels, the production intermittency requires more extended use of storage, both for heat and electricity regeneration. The particular example in Figure 2.6 handles the intermittency of wind power by international grid import–export options, such as those available in the Scandinavian network, based on large hydro reservoirs in Sweden and particularly in Norway. It is important to note that no pumping of water upwards to the reservoirs is needed because the "repayment" of power by the renewable energy system is made by displacing hydro generation at times of wind surplus, and hydro coverage of the wind customers in case of insufficient wind is only affecting the power rating of the turbines (Sørensen 1981).

References

Jensen, J., and B. Sørensen. 1984. *Fundamentals of energy storage*. New York: Wiley-Interscience.

OECD/IEA. 2005. *Learning from the blackouts*. Energy market experience series. Paris: International Energy Agency.

Sørensen, B. 1981. A combined wind and hydro power system. *Energy Policy* 9 (1): 51–55.

Sørensen, B. 1983. Stationary applications of fuel cells. In *Solid state protonic conductors II for fuel cells and sensors*, ed. J. Goodenough, J. Jensen, and M. Kleitz, 97–108. Odense, Denmark: Odense University Press.

Sørensen, B. 1984. Energy storage. *Annual Review of Energy* 9:1–29.

3

TIMESCALES RELEVANT FOR THE INTERMITTENCY OF INDIVIDUAL ENERGY SOURCES

Variations in availability of energy flows from different energy sources have components on timescales ranging from seconds to years, resulting from short-term fluctuations in cloud cover or wind turbulence, from diurnal effects of the rotating Earth, over seasonal variation, and to interannual differences caused by climatic anomalies.

For fuel-based energy systems, most intermittency problems are caused by system failures and are thus difficult to predict. Serious prolonged supply interruption may happen, e.g., for natural gas or oil supply by long pipelines, which may require extended repair times for certain types of disruptions, such as caused by undersea line damage by fishing trawls or by submarines severing gas pipes. For nuclear power stations, prolonged disruptions have happened as a result of major reactor accidents as well as in cases where unanticipated safety problems not previously identified become apparent, leading to a long-time ban of operation by the controlling authorities for several reactors of similar construction. Because of the nature of such causes of intermittency, their backup requirements are difficult to ascertain except by analysis of historical experience. Nuclear accidents can cause the loss of many gigawatts of generating capacity for over a decade because of the long construction time for new nuclear facilities and the need to improve design relative to the one that failed before new construction is attempted. Regulatory closure of nuclear plants has in the past lasted for periods of about a year before new procedures or hardware amendments could, respectively, be implemented or installed. Most natural gas and oil pipeline failures have been repaired within some months, but cases of large undersea disruptions do not have sufficient statistical data to be meaningfully subjected to stochastic analysis.

For renewable energy supply systems, the data are generally available. Solar and wind generating facilities currently have unit sizes that, even for the largest wind turbines, do not reach 10 MW, compared to the gigawatt rating of large nuclear or hydro facilities. Due to this decentralized structure of most solar and wind generation facilities, intermittency from failure of individual units can usually be neglected due to the stochastic distribution underlying such failures, and the intermittency will thus predominantly be due to the fluctuations in resource flows, which can be fairly accurately predicted by meteorological modeling. The implication is that the remedies for dealing with intermittency can be identified as part of establishing the supply systems, and the operational stability of renewable energy systems containing specific solutions to the known intermittency problems is therefore higher than for systems depending on large generating units of partially unknown failure probabilities. Exceptions to this conclusion could occur for design flaws affecting a large number of identical renewable-energy conversion units.

Technically, the relevant parameter is the *loss of load probability* (LOLP), which can be estimated if there is sufficient data available (Haslett and Diesendorf 2011). For electric utility systems in Europe, the LOLP deemed acceptable is typically up to 10^{-4} (about one hour inability to satisfy load over a year), and values below 10^{-5} are rarely reached. In several other parts of the world, the actual LOLP is higher.

Time series of natural variations in the source flows of various energy types used in energy supply systems can be instructive for estimating storage requirements or assessing other measures that will take care of the fluctuation or intermittency. However, as mentioned previously, not only resource flow, but also load variations and system failures contribute to the supply–demand mismatch, whether expressed in terms of the LOLP or otherwise.

For fuel sources such as nuclear, fossil, or wood fuels, the time sequences of extraction is fairly independent of the further uses of the energy provided, as the high energy density makes storage of these fuels rather inexpensive. An exception may be natural gas, often produced in conjunction with oil. While oil products may be transported by vehicles across sea or land, a gaseous fuel must first be compressed or liquefied to allow such simple transport. Otherwise, pipelines are used for quite long distances over land or at sea, but of course augmenting the cost. In fact, until recently (and still practiced in some backward countries), natural gas was often flared at the site of production, a practice associated with unnecessary CO_2 emissions, or was reinjected into the boreholes without avoiding releases of methane, a very potent greenhouse gas (Tollefson 2013). Inferior fossil energy sources such as tar sands or oil shale have long been identified but are currently beginning to be exploited. Because of the extreme environmental impact of their extraction and treatment, often making the mining area unsuited for human occupation for long periods of time, such mining operations are of doubtful acceptability for general energy supply. The negative impacts may not be more severe than for other mining operations, but the areas affected are often much larger than for deep mineral mining of energy raw materials and comparable to those for the surface extraction of coal that currently takes place in remote, uninhabited areas (for example Australia, where most inhabitants do not seem to particularly value wilderness areas). However, the storage and transport impacts are no different from those of fossil fuels from richer sources, provided that the separation and refining is done at the remote site of extraction. Currently, storage is in effect not only for solid fuels, but also for natural gas (in underground caverns or in building-size containers, ideally—but not always—placed away from inhabited areas for safety reasons) to better match demand variations.

Biomass resources have many uses: for food, for construction materials (e.g., wood poles, planks, and furniture), and as feedstock for chemical and other industries (for glues, pharmaceutical goods, and various additives), in addition to energy uses (combustion, biogas, and advanced biofuels production). For short-rotation biomass crops, there is a time-dependence associated with harvesting times. These may be important for the management of further conversion, both for food and for energy raw materials. The price currently paid for food is generally an order of magnitude higher than the present cost of fuels, when compared on the basis of energy content, which is why there is a market for long-distance transport of food items from areas of recent harvest to regions where the particular item is not in season. For energy use, the additional cost of long-distance transport may not be warranted, and the seasonal variations may therefore be a constraining factor. However, if the biomass is converted into a fuel that can be stored (e.g., a liquid biofuel), the issue may not be important.

In order to obtain an overview, one can look at the seasonal variations in standing crop biomass, which is illustrated on a global scale in Figures 3.1a–l, based on satellite observations by the US Geological Survey NASA Land Processes Distributed Active Archive Center (NASA 2013). The parameter shown, Enhanced Vegetation Index (EVI), is derived from surface reflectivity at different wavelengths of the solar radiation spectrum, using specifically the reflectance at the three wavelengths 800 nm, 680 nm, and 450 nm to exhibit the changes over the year in leaf chlorophyll concentration. A simpler such index of the extent of green vegetation is the Normalized Difference Vegetation Index (NDVI) defined by the reflectance R at only two wavelengths,

$$\text{NDVI} = (R_{800} - R_{680})/(R_{800} + R_{680}) \tag{3.1}$$

The EVI attempts to correct for soil background and atmospheric influences by invoking the blue light reflection in an ad hoc way (Huete et al. 1997),

$$\text{EVI} = 2.5 \, (R_{800} - R_{680})/(R_{800} + 6 \, R_{680} - 7.5 \, R_{450} + 1) \tag{3.2}$$

Over water, EVI has a value of about –0.3, and for perfect snow cover a value of zero, while the value for land areas with cities, roads, rocky surfaces, and various types of vegetation ranges from zero to one.

What may strike an observer viewing the seasonal variations in vegetation cover in Figure 3.1 is how little variation there is over the year at any particular region of the world. General determinants of plant growth are solar radiation, precipitation, soil type, and moisture, plus the effects of agricultural management, including crop choices, plowing, tillage work, and possibly irrigation. The geographical variations are large, with many desert areas having little vegetation, but variability in solar radiation seems less important, as substantial plant cover is found at quite high latitudes, although there is an effect at the very highest latitudes, due to low temperatures and seasonal absence of solar radiation. But at slightly lower latitudes, the longer days during the summer growing season are seen to counteract the lower solar radiation. Clearly, because the vegetation cover lasts longer than the period of plant growth, the vegetation index is not a good measure of productivity. Plant growth may to some extent be derived by comparing diurnal satellite data, but in most cases, treatment in terms of analytical models will increase the quality of results (cf. Chapters 3 and 6 in Sørensen [2010]).

Figures 3.1a–c show that from January to March, the high plant cover below the equator, notably in Africa and South America, is reduced, but there is no compensating growth in greenness north

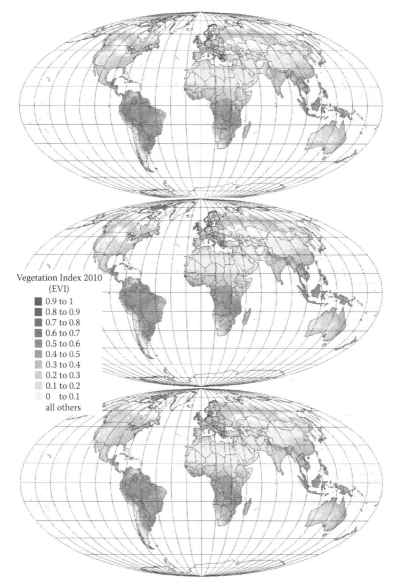

Figures 3.1   (a–c) Index of vegetation cover (EVI, defined by Equation [3.2]) for January (top), February (middle), and March (bottom) in 2010, based on satellite data from NASA (2013).

Continued

of the equator. The following three months, April to June shown in
Figures 3.1d–f, show a continued weakening of the vegetation cover
on the Southern Hemisphere, while that of the Northern Hemisphere
increases in area and reaches far northern areas of Siberia, Canada,
and Scandinavia. Vegetation coverage in central Europe, along with

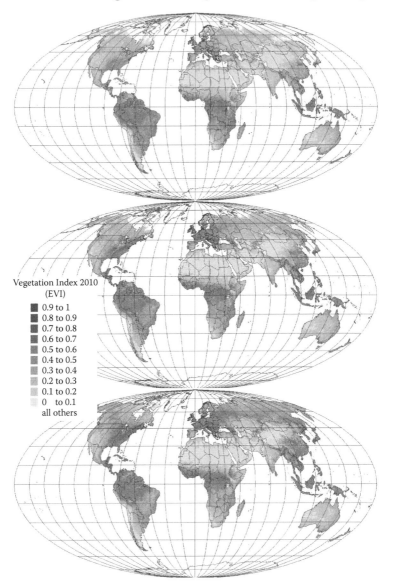

Figures 3.1 (*Continued*)   (d–f) Index of vegetation cover (EVI, defined by Equation [3.2]) for
April (top), May (middle), and June (bottom) in 2010, based on satellite data from NASA (2013).

Continued

that of the eastern United States and to some extent eastern China, increases strongly already in May.

Over the months of July to September, the vegetation cover in the northern agricultural regions drops progressively, as the main harvest is removed from the land (Figures 3.1g–i). Also in the Southern

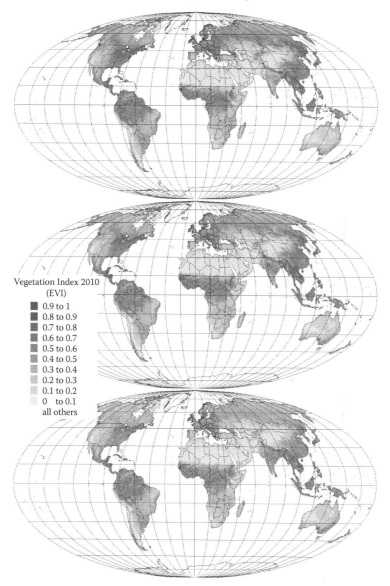

Figures 3.1 (*Continued*)   (g–i) Index of vegetation cover (EVI, defined by Equation [3.2]) for July (top), August (middle), and September (bottom) in 2010, based on satellite data from NASA (2013).   *Continued*

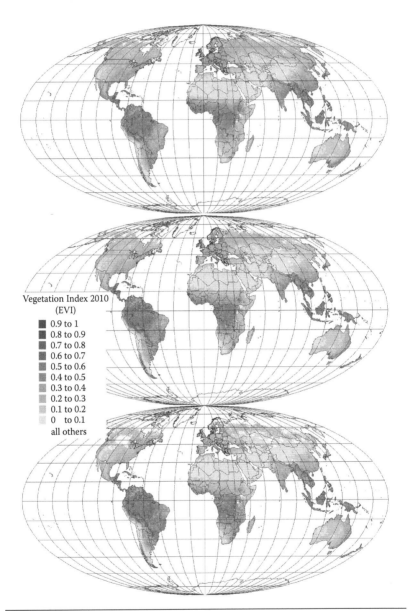

Figures 3.1 (Continued)   (j–l) Index of vegetation cover (EVI, defined by Equation [3.2]) for October (top), November (middle), and December (bottom) in 2010, based on satellite data from NASA (2013).

Hemisphere, the plant cover is gradually reduced over this period. This may have to do with agricultural practices, where the number of crops grown in succession varies. Near the equator, more than one crop annually is the standard, and sowing times are often determined by the occurrence of a rainy season. Due to soil erosion, the land cannot be left without a crop for long. In northern Europe, the tradition of a single crop prevailed until a few decades ago, and the high soil moisture largely eliminated the erosion problem. Today, in most of northern Europe, a second crop (often called the "winter crop") is becoming standard. It is often a high-protein crop like alfalfa (also called lucerne) replacing the imported protein-boosters used earlier. Despite the name *winter crop*, the short growing season for this crop is most often late fall or spring, thereby avoiding the winter months with risk of snow cover.

Finally, the EVI for the months of October to December, shown in Figure 3.1j–l, shows declining values in the Northern Hemisphere but little altered ones in the southern regions with forest or agriculture. EVI for the most northern regions in Europe, Asia, and North America drops to zero, as soils become barren or snow covered. In the Southern Hemisphere, factors other than solar radiation are important, and in any case, the inhabited areas are not nearly as far south as the corresponding ones on the Northern Hemisphere are far north. For example, vegetation on the southeastern coast of Australia is highest in the period August to September and low during the southern summer months, which are also drier.

Considering solar radiation, which can be directly collected by solar thermal or solar electrical panels or dishes, the variations over the year may similarly be estimated from satellite data, pertaining to horizontal surfaces and taking into account the attenuation by clouds and atmospheric particle content (ECMWF 2008). In order to estimate energy production, it is also necessary to consider tilted surfaces facing the sky in different directions (as discussed in Sørensen [2010, chap. 3]), but an overview can still be gained by just looking at the seasonal variations of radiation on a horizontal surface.

The January solar radiation (Figure 3.2a) is very high on the Southern Hemisphere, both at low and high latitudes, and it diminishes in February and March (Figures 3.2b,c), particularly near the South Pole. On the Northern Hemisphere, there is in January very little solar radiation above 30° N, and it only increases modestly over the next months. From April to June (Figures 3.2d–f), high southern latitudes lose the radiation while the corresponding northern latitudes gain solar radiation. Particularly in May and June, the radiation received in Greenland is very high, caused by the combination of long days and low atmospheric particle content (compared with the other high-latitude areas in Canada and Siberia). Continuing to the period July to September (Figures 3.2g–i), it is seen that there is not complete symmetry between spring and autumn. In autumn, haze near the equator causes decreased solar radiation, and the reduction of radiation at high northern latitudes in September is smaller than the corresponding one at southern latitudes during April. The reason is scattered (i.e., nondirect) radiation over the northern continental land areas (with higher temperatures and more dust) as compared to the corresponding southern regions that are dominated by ocean waters. Figures 3.2j–l show the solar radiation on a horizontal plane for the final months of year 2000. Note that land areas, and particularly those on eastern rims of continents, receive much more radiation than ocean areas at the same latitudes, and that the radiation in December on the Northern Hemisphere is even lower than in January. These features are consequences of the full set of climate determinants, including temperatures in the atmosphere as well as at the land or sea surfaces, moisture, winds, particle content, and clouds, all influencing the disposition of incoming solar radiation (Sørensen 2010).

An important renewable energy resource is hydro energy, traditionally used for mechanical energy by means of waterwheels but currently mostly utilized in electric power generation mode, using running water (rivers, streams) and elevated water reservoirs (lakes, artificially created water basins). The basis for hydro energy is the water cycle, where water evaporated from land or ocean is returned by precipitation (rain, snow), and if this is over land, then at the elevation of the site and thus constituting potential energy. Some of the water may reevaporate and some may contribute to the hydrological cycle by adding soil moisture or joining groundwater reservoirs, but any surplus will be available for runoff at the surface. Considerable delays between precipitation and joining meltwater streams may occur for snow and ice at high elevation or high-latitude locations.

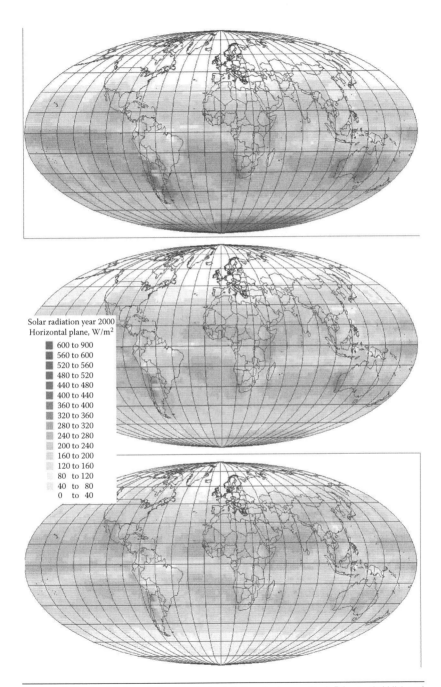

Figures 3.2   (a–c) Solar radiation on a horizontal plane for January (top), February (middle), and March (bottom) in 2000, based on satellite data from ECMWF (2008).   *Continued*

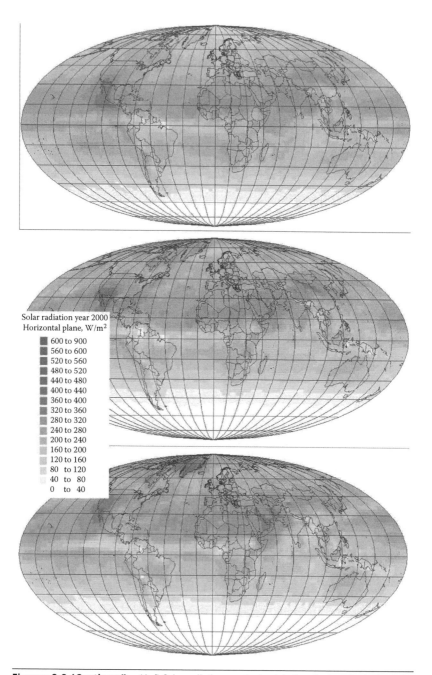

Figures 3.2 (*Continued*)   (d–f) Solar radiation on a horizontal plane for April (top), May (middle), and June (bottom) in 2000, based on satellite data from ECMWF (2008).   *Continued*

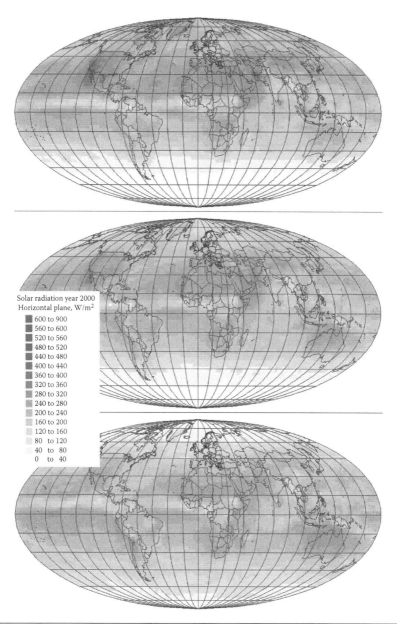

Figures 3.2 (*Continued*)   (g–i) Solar radiation on a horizontal plane for July (top), August (middle), and September (bottom) in 2000, based on satellite data from ECMWF (2008).

Continued

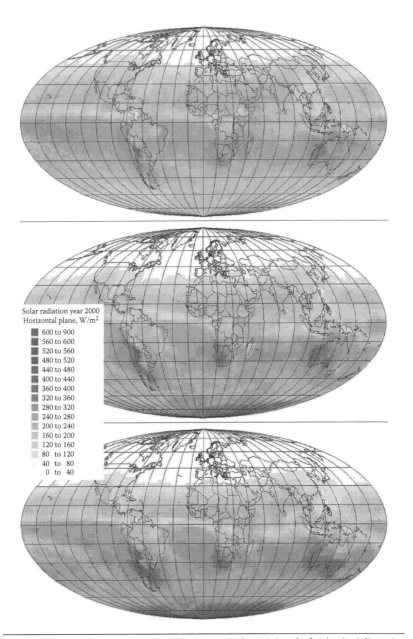

Solar radiation year 2000
Horizontal plane, W/m^2

■ 600 to 900
■ 560 to 600
■ 520 to 560
■ 480 to 520
■ 440 to 480
■ 400 to 440
■ 360 to 400
■ 320 to 360
■ 280 to 320
■ 240 to 280
■ 200 to 240
■ 160 to 200
■ 120 to 160
■ 80 to 120
■ 40 to 80
■   0 to 40

Figures 3.2 (*Continued*)   (j–l) Solar radiation on a horizontal plane for October (top), November (middle), and December (bottom) in 2000, based on satellite data from ECMWF (2008).

Surface runoff is thus any water flowing along the land surfaces. Subtleties include groundwater later joining the surface flow at a location different from the area of entry. The surplus water may form pools or add to lake water levels before joining streams and rivers and making its way toward the sea. Figures 3.3a–l (NOAA 2013) show the potential hydropower based on monthly surface runoff (r, measured in kilograms of water per square meter of land and per second). Multiplying by the elevation (h, unit meter) and the gravitational constant, which at the surface of the Earth is $g = 9.81$ m/s^2, one gets the potential hydropower P in watts per square meter (W/m^2),

$$P = r g h \tag{3.3}$$

This potential power displays seasonal effects as well as other peculiarities of the atmospheric processes, including equatorial monsoon effects and general temperature dependence as well as the precipitation enhancement happening when moist air hits mountain regions. For South America, the maximum is found in January, with strong runoff also in February to April. The flow activity diminishes until the end of the year. In Africa, there are two maxima or "rainy seasons," January to March and August to October. In the Tibet Highland, runoff is strong all year, but with some increase in January and July. For northern North America, northern Europe, and Siberia, a high level of potential hydro energy is found in the winter season December to April, but it reaches very low values during the northern summer. The regions near the poles have large potential hydro all year. It is important to note that the energy flows associated with runoff are only "potential," as two conditions must be fulfilled to realize the option for making use of the hydropower: The surface water must assemble into an elevated reservoir with suitable outlet (e.g., a dam separating the reservoir from a lower waterway), a stream, or a river in order for hydro turbines to be installed, and the water must be liquid in order to drive a turbine. In cold climates (such as Norway, Canada, Siberia), the water surplus available as snow or ice can only be used when it melts, which may be several months after being registered as "runoff" in the frames of Figure 3.3. Naturally, water precipitated at low heights does not offer as much potential for energy extraction as precipitation falling at higher elevations.

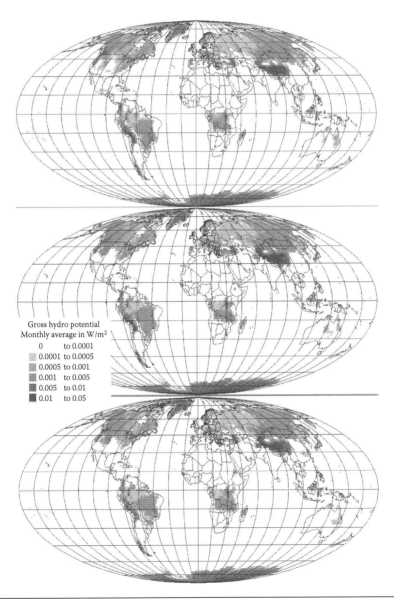

Figures 3.3   (a–c) Potential hydro energy (in W/m²) for January (top), February (middle), and March (bottom) of 2010, based on data from NOAA (2013). A few areas miss data, notably islands like Tasmania.   *Continued*

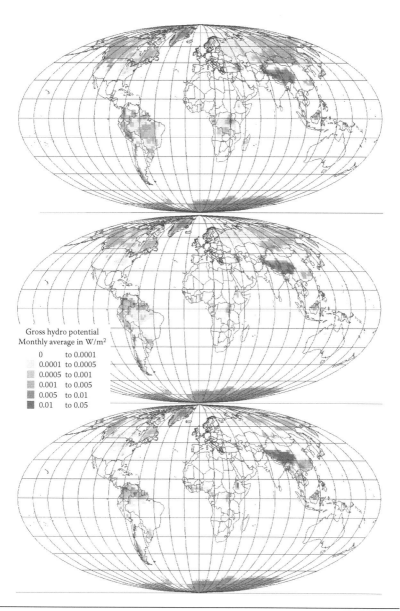

Gross hydro potential
Monthly average in W/m²

0	to 0.0001
0.0001	to 0.0005
0.0005	to 0.001
0.001	to 0.005
0.005	to 0.01
0.01	to 0.05

Figures 3.3 (*Continued*)   (d–f) Potential hydro energy (in W/m²) for April (top), May (middle), and June (bottom) of 2010, based on data from NOAA (2013). A few areas miss data, notably islands like Tasmania.   *Continued*

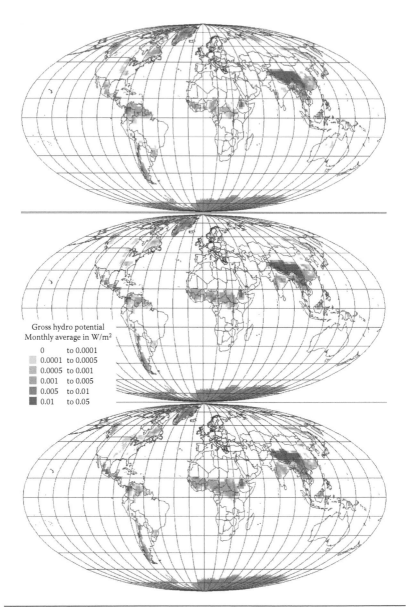

Figures 3.3 (*Continued*)   (g–i) Potential hydro energy (in W/m²) for July (top), August (middle), and September (bottom) of 2010, based on data from NOAA (2013). A few areas miss data, notably islands like Tasmania.   *Continued*

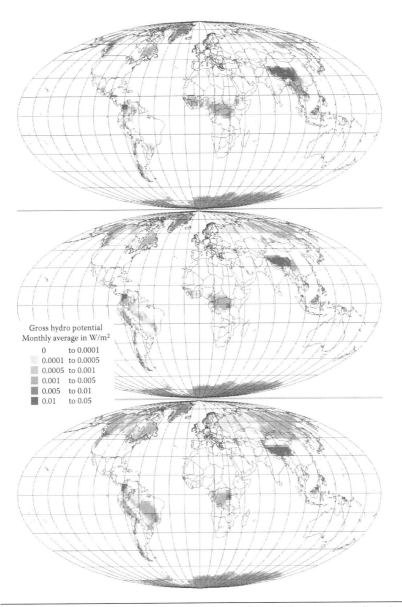

Figures 3.3 (*Continued*)   (j–l) Potential hydro energy (in W/m²) for October (top), November (middle), and December (bottom) of 2010, based on data from NOAA (2013). A few areas miss data, notably islands like Tasmania.

It should also be noted that the variations in flow exhibited in Figures 3.3a–l are most important for run-of-the-river hydro plants, i.e., plants where part of the water in a river is led through hydro turbines. If the scheme is using high-lying reservoirs, these act like energy stores and allow the extraction of power to be made independent of the water cycle filling the reservoirs. The hydro options available in Canada, northern Scandinavia, and Siberia often allow annual energy storage, as the surplus water level caused by snowmelt in spring and early summer may last an entire year, until the next melting period. It should also be kept in mind that, if forming the reservoirs to hold the water until it is needed for power production is done by stemming up water into the reservoirs by large dam construction, then there will be substantial environmental impacts from the hydro energy scheme as well as social impacts, if people were already living in the areas flooded to form the reservoirs (Sørensen 2011).

For wind energy, the seasonal variations are depicted in Figures 3.4a–l, based on data for year 2000 (NCAR 2006). Generally speaking, wind power levels are higher over water, where the surface roughness slowing the wind flow is tiny. However, offshore wind turbines need foundation, which is currently feasible to depths of about 50 m, or they should be mounted on floating platforms, a technology not fully developed at present. The general impression is that seasonal variations are small, but focusing on onshore or near-shore locations, the variations are larger, although not nearly as large as those of, for example, direct solar energy. The horizontal energy flows in the wind have been folded with the power characteristics of typical wind turbines, producing no power at wind speeds below about 5 m/s, reaching the maximum efficiency in the range of 12–15 m/s, and flattening out above some 20 m/s, indicating a declining flow-to-power conversion efficiency.

The January wind potential (Figure 3.4a) is very high at the North American east coast, northern Europe (the British Isles, Iceland, and the Nordic countries), eastern Mediterranean (Greek Islands and Turkish west coast), the Asian east coast and Japan, and reasonably high in Chile, South Africa, and both the western and eastern coasts of Australia. The southwest African coast and the western Australian coast have higher winds in February (Figure 3.4b), whereas the Northern Hemisphere winds decline a bit. This is also true for March

(Figure 3.4c), but the British Isles and Denmark retain high winds, as do the North American lake areas. Winds on the African and South American continents decline, except for the southwestern African coast, but in Australia, winds remain high on the west coast as well as on Queensland coastal regions.

In April (Figure 3.4d), northern Europe, the Chinese east coast, and Japan retain high wind potentials, and winds at the U.S. east coast increase, while the southern sites have declining winds. Strong winds associated with equatorial circulation appear on the eastern coasts of Central America, the western coasts of Africa, and the east coast of India. However, these disappear already in May (Figure 3.4e), where the northern European winds also decline. However, winds on the eastern Central America coasts remain sizeable, and in June (Figure 3.4f), both Indian coasts exhibit high winds, along with the equatorial east coast of Africa and the south shores of Indonesia along with the north shores of Australia. This pattern is reinforced in July (Figure 3.4g), but now also the continents in the Southern Hemisphere all have substantial wind potential. In the following months, the southern winds decline, and by September (Figure 3.4i), the equatorial regions have little wind potential, while that of northern Europe and northwest Africa increases. Wind potentials in India remain high in August (Figure 3.4h) but drop in September. By October (Figure 3.4j), the wind potential in Japan and the eastern Chinese coasts again become large, as do those of both the west and east coasts of North America. The wind potentials in the British Isles, Iceland, and the Nordic countries become very high, as do winds in Central America (east coasts), northern Brazil as well as the South American west and east coasts and the African west and east coasts, except for the equatorial region. Australia and New Zealand also see fine winds, highest at the west and south coasts. These patterns prevail for November and December (Figures 3.4k,l), although the wind potential declines somewhat, and Antarctic winds cease for the first time during the year.

This concludes the summary of the seasonal variations of the major renewable energy sources, based on data for a particular year. There are variations between years that may shift some of the patterns, but the overall picture remains approximately as shown in Figures 3.1–3.4. As mentioned in the beginning of this chapter, the renewable energy flows

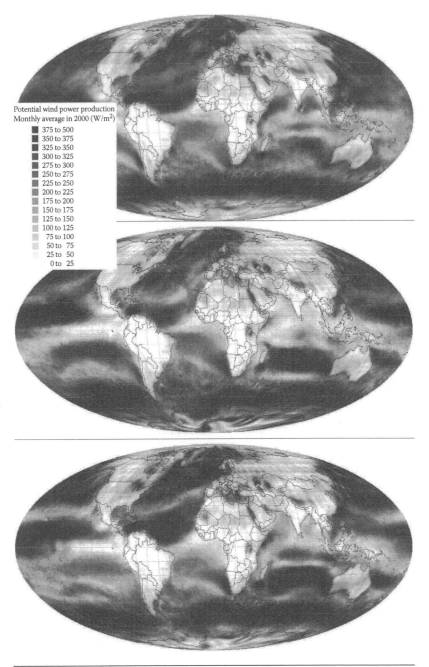

Potential wind power production
Monthly average in 2000 (W/m²)

- 375 to 500
- 350 to 375
- 325 to 350
- 300 to 325
- 275 to 300
- 250 to 275
- 225 to 250
- 200 to 225
- 175 to 200
- 150 to 175
- 125 to 150
- 100 to 125
- 75 to 100
- 50 to 75
- 25 to 50
- 0 to 25

Figures 3.4 (a–c) Potential wind power production (in W/m²) for January (top), February (middle), and March (bottom) of 2000, based on data from NCAR (2006) (used with permission.), folded with a power curve of typical wind turbines (Sørensen 2008). *Continued*

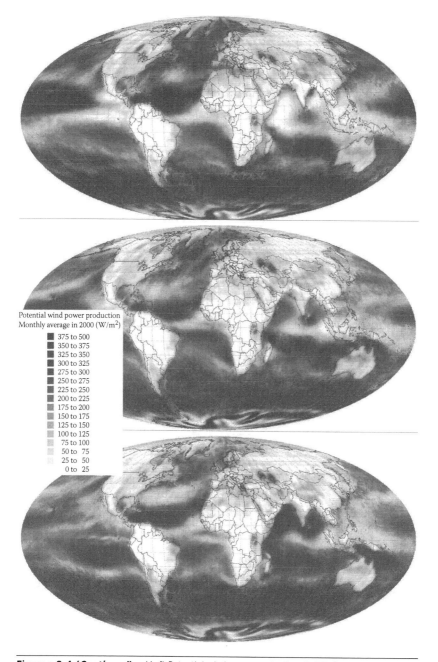

Figures 3.4 (*Continued*)   (d–f) Potential wind power production (in W/m²) for April (top), May (middle), and June (bottom) of 2000, based on data from NCAR (2006) (used with permission.), folded with a power curve of typical wind turbines (Sørensen 2008).   *Continued*

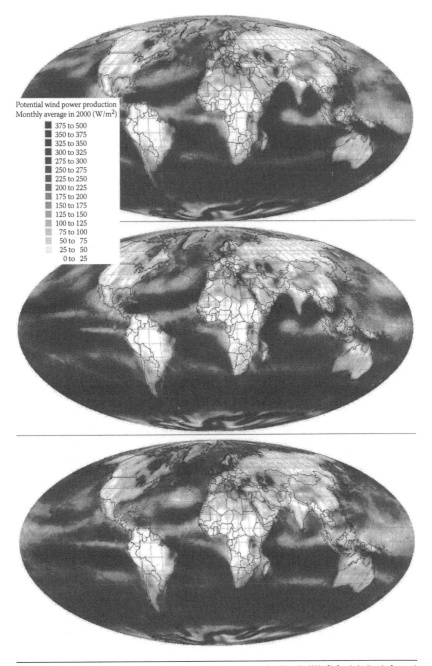

Figures 3.4 (*Continued*)   (g–i) Potential wind power production (in W/m²) for July (top), August (middle), and September (bottom) of 2000, based on data from NCAR (2006) (used with permission.), folded with a power curve of typical wind turbines (Sørensen 2008).   *Continued*

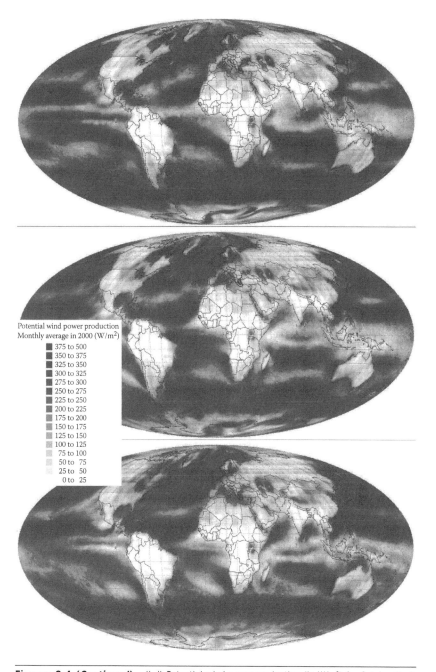

Figures 3.4 (*Continued*)   (j–l) Potential wind power production (in W/m²) for October (top), November (middle), and December (bottom) of 2000, based on data from NCAR (2006) (used with permission.), folded with a power curve of typical wind turbines (Sørensen 2008).

also exhibit variations on shorter timescales, such as the day-to-night variations in solar energy and the effect on solar radiation at the Earth's surface from cloud passage, the dependence of wind power on the passage of particular weather pressure systems over a particular region, and the dependence of hydro energy without reservoirs (based on the run of a river) on the pattern of rainfall. Very short-term variations, such as the chaotic fluctuations in wind energy at a specific location, rarely affect energy systems, due to the modular nature of renewable energy systems, often deriving power from a substantial number of individual units and thus smoothing out the variability on the scale of seconds.

For nonrenewable energy sources, variability is as mentioned due to less predictable factors, such as component failures and accidents, and timewise patterns are known only in a statistical sense based on past experience, for which some collections exist, for example, as discussed by Sovacool (2008).

References

ECMWF. 2008. European Centre for Medium-Range Weather Forecasts 40 Years Reanalysis (ERA40). http://data-portal.ecmwf.int.

Haslett, J., and M. Diesendorf. 2011. The capacity credit of wind power: A theoretical analysis. In *Renewable energy reference collection*. Vol. 4, ed. B. Sørensen, 17–36. London: Earthscan.

Huete, A., H. Liu, K. Batchily, and van Leeuwen. 1997. A comparison of vegetation indices over a global set of TM images for EOS-MODIS. *Remote Sensing of the Environment* 59 (3): 440–51.

Kalney, E., M. Kanamitsu, R. Kistler, W. Collins, D. Deaven, L. Gandin, M. Iredell, et al. 1996. The NCEP/NCAR 40-year reanalysis project. *Bulletin American Meteorological Society* 77 (3): 437–71. http://rda.ucar.edu/datasets/ds090.0/docs/publications/bams1996mar/bams1996mar.pdf.

Milliff, R., J. Morzel, D. Chelton, and M. Freilich. 2004. Wind stress curl and wind stress divergence biases from rain effects on QSCAT surface wind retrievals. *J. Atmos. Ocean. Tech.* 21:1216–31.

NASA. 2013. Land Processes Distributed Active Archive Center (LP DAAC), USGS/Earth Resources Observation and Science (EROS) Center, Sioux Falls, SD. Online Data Pool: MODIS/AQUA Vegetation Indices MYD13C2 for the year 2010. https://lpdaac.usgs.gov.

NCAR. 2006. QSCAT/NCEP blended ocean winds from Colorado Research Associates, ds744.4. Research Data Archive at National Center for Atmospheric Research. http://rda.ucar.edu/datasets/ds744.4. (The methods of collecting and using the data for wind studies are described in Milliff et al. [2004] and Sørensen [2008].)

NOAA. 2013. NCEP-NCAR CDAS-1 monthly diagnostic surface runoff data. (General description is provided in Kalney et al. [1996].) Available from the Columbia University data collection at http://iridl.ldeo.columbia.edu/SOURCES/.NOAA/.NCEP-NCAR/.CDAS-1/.

Sørensen, B. 2008. A new method for estimating off-shore wind potentials. *International Journal of Green Energy* 5:139–47.

Sørensen, B. 2010. *Renewable energy: Physics, engineering, environmental impacts, economics and planning.* 4th ed. Burlington, MA: Elsevier.

Sørensen, B. 2011. *Life-cycle analysis of energy systems: From methodology to applications.* Cambridge, UK: RCS Publishing.

Sovacool, B. 2008. The cost of failure: A preliminary assessment of major energy accidents, 1907–2007. *Energy Policy* 36:1802–20.

Tollefson, J. 2013. Oil boom raises burning issues. *Nature* 495 (7441): 290–91.

4

USING CASE STUDIES TO EXPLORE THE OPTIONS

In this and the following chapters, many of the intermittency issues will be illustrated by means of simple simulations of actual systems. In these examples, the systems are often simplified in such a way that the immense complexity of real-world systems is replaced by a small number of idealized system components capable of singling out the behavior relevant to the particular ways of dealing with demand–supply mismatch by storage, energy trade, or demand management. These simplified systems are also offered as basic scenarios for the structure of future energy systems. The attraction of looking a bit ahead is that a future system can be constructed in the "right" way from the beginning, without the extra cost that would occur if an existing system were to be modified or replaced before having reached the economic life period expected when the system was built. One such reference scenario is presented in Section 4.1 for North America, starting with the contiguous part of the United States and then adding the rest of North American nations or regions in order to later explore options for energy exchange and trade. Other scenarios for future energy systems will be used to illustrate issues in the following chapters, and a final Asian scenario is subjected to simulation toward the end of the book.

Quite generally, energy systems may be assessed at various stages along the traversed energy-transformation paths. Energy sources are transformed into primary energy inputs with some inherent losses (e.g., energy spent moving oil to the surface and refining it into a variety of light and heavy components). Then further conversion takes place in order to bring the energy to the consumer in the desired form (e.g., by the Carnot processes* of fossil-fueled power plants), and

---

* Performed by any plant using thermodynamical cycles to convert heat from fuels into electricity (and possibly making use of associated heat). See list of cycles in Chapter 4 of Sørensen (2010).

finally, the energy delivered to the final user may have to be transformed in order to satisfy the particular service wanted (such as food preservation by refrigerators or overcoming road friction and elevation differences by driving automobiles). The conversions taking place at the end user define the final end-use energy demand. Most national statistics confine themselves to giving the energy delivered to the end user, i.e., before his or her final conversion. A detailed description of the considerations leading to different scenarios for the magnitude of end-use energy demand in the future may be found in Sørensen (2008a), along with the conversions between delivered energy and the ultimate end-use energy. Because of the familiarity in statistical data sources with the delivered energy demands, the following examples will deal with demand in terms of the energy delivered to the final user by drawing upon the relations between the different assessment points defined by Sørensen (2008a).

The energy sources exploited in the scenarios here are the most important renewable energy resources. Because the scenarios are only intended as schematic, the small contributions from, e.g., geothermal or wave energy are omitted. This leaves wind, direct solar radiation, hydropower, and bioenergy, the general availability of which was examined in Chapter 3. The detailed working of the devices used to capture and convert these energy sources are not the subject of this book, but further details may be found in engineering textbooks such as Sørensen (2010). Figures 4.1–4.10 give an idea of the kind of devices employed in the scenarios. Figure 4.1 shows land-based wind turbines placed on agricultural land. Larger arrays of such machines may be (and are, in some locations) placed on marginal land. The restrictions on other uses of the land are very modest, and should better energy solutions emerge in the future, the wind turbines may be removed without having made any lasting impact on the environment. Current turbines have small noise impacts, and the visual impacts are seen as pleasant by farmers (having used wind energy for at least 500 years) and by people realizing that they are designs made with architectural care similar to that of church towers or other visually intrusive accompaniments of human belief sets, but they are seen as ugly by observers employed in or favoring competing industries such as the nuclear one. The wind turbines erected on windy locations already deliver power at costs similar to current alternatives (such as coal-based Carnot-cycle power plants).

Figure 4.1 Wind turbines (manufactured by Vestas) on agricultural land in Denmark (photo by author). The impact on farming land use is minimal, allowing plant growth, tillage, and plowing a few meters from the towers (Sørensen 2012b, Fig. 10.20; used with permission).

Offshore wind turbines offer a very attractive way of augmenting wind-power production for countries endowed with suitable potentials. The higher costs, due to the construction of foundations and slightly more expensive maintenance work (due to access requiring boats), are on average compensated for by the higher production at sea than on land for the same turbine size. Due to the distance from habitations, the impacts (noise and visual) are even lower than on land (see Figure 4.2).

Use of hydropower in the scenarios is largely restricted to those installations already in operation or under construction. The implied water-turbine technologies have been around for a very long time. Large hydropower facilities can have severe environmental effects, and there are examples of construction of hydropower dams that

Figure 4.2 Offshore windmill park at Hornsrev, Denmark. The nearest of the 2-MW turbines are about 15 km from the shore (video still by author; the blur affecting the turbines is also due to haze and atmospheric turbulence) (Sørensen 2012b, Fig. 10.21a; used with permission).

have displaced large numbers of people who used to live in the area. Smaller run-of-the-river installations can usually be made with modest negative impacts. The discussion of hydro reservoirs is continued in Chapter 8.

Solar radiation can be used directly in several ways, in addition to the indirect ways by growing plants and trees or interfering with the Earth's rotation to create winds. In deserts with a high proportion of direct radiation (i.e., not scattered and thus coming from the direction of the Sun within a very small margin), mirrors can be used to reflect and concentrate the radiation on a small receiver, the temperature of which may rise to several thousand degrees and drive a Carnot-cycle power plant, producing electricity at a high efficiency (Figure 4.3). In other areas with a fair percentage of direct solar radiation, a tracking photovoltaic device such as the one shown in Figure 4.4 can augment the power production per unit collector area. Under the circumstances prevailing in most remaining areas (constituting the largest proportion of all areas suitable for solar collection), there are substantial amounts

Figure 4.3   Concentrating solar thermal-electric installation in Mojave Desert, California. This type of solar collector requires a high fraction of the solar radiation as direct, which is typical of deserts but few other places (photo by author).

Figure 4.4   Tracking photovoltaic collector in Saudi Arabia (photo by author).

of scattered radiation from clouds, from particles in the atmosphere, and from turbulence of the air. Here a flat-plate, fixed solar collector mounted at the best average tilt angle toward the south will often suffice, leaving only small efficiency improvements to be achieved by the fairly costly tracking or concentrating methods. Figure 4.5 shows such an installation in the form of a large-area solar panel field placed on marginal land. The same type of photovoltaic panels can be

Figure 4.5   Centralized photovoltaic farm (400-kW amorphous cells) placed on marginal land at Davis in California as part of the PVUSA Utility Scale Systems program (photo by author).

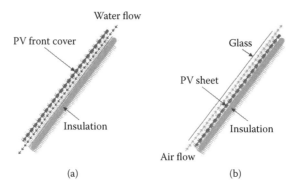

Figure 4.6 Typical designs of photovoltaic-thermal (PVT) collector panels. Heat is extracted to (a) a water flow or (b) an air flow. The water channel in (a) is below the PV electricity-collecting sheet, while the air channel in (b) is above, necessitating a cover of glass or another transparent material.

used on a small scale for building-integrated installations, covering a south-facing roof or façade. In many of the scenarios, heat will also be derived from the building-integrated solar panels, assumed to be of the PVT-type, where the same device produces power and heat (see the review by Chow [2010]). This is needed because of the limited area available on buildings, and it is advantageous due to the relatively low efficiency of photovoltaic conversion, which for current solar panels is in the range of 10%–20%. That leaves considerable amounts of "waste heat," of which some 50% can be extracted by suitable water- or air-based piping (Figure 4.6), connected to a thermal store placed in the building. The efficiency variation of the thermal collection is far more complex than that of the electricity collection, because it depends on the temperature of the store. Several examples of the behavior that can be anticipated are discussed in Sørensen (2010).

A variety of biofuels can be produced with agricultural or forestry biomass residues as feedstock, or with use of biomass such as kelp produced by aquaculture. They will enter most of the scenarios as fuels for the transportation sector (Figure 4.7), in competition with electric and fuel-cell vehicles until they are eventually phased out when alternatives with lesser environmental impacts have become more viable than they are today. Most of the scenarios in this book use a considerable part of the available biomass residues for transportation fuels, but do invoke electric-powered (Figure 4.9) or hydrogen-powered vehicles (Figure 4.8) if there is insufficient biofuels, thus assuming

Figure 4.7 Biofuel car operating in Finland (photo by author).

Figure 4.8 One of several prototype hydrogen fuel-cell cars built over the past 20 years (Necar-5 by Daimler AG; photo by author).

that the scenario realization is so far into the future (2050 or 2060 in the actual scenarios constructed) that zero-emission alternatives have become available and viable. The most likely such technology for road transportation is a hybrid fuel-cell–electric vehicle (Sørensen 2012a). For tracked transportation, electricity is already today the preferred energy solution, as seen in Figure 4.10 for a monorail, and is actually

Figure 4.9   Charging of electric car (Smart for Two in Amsterdam; photo by L. Hirlimann, Wikimedia Commons public domain).

Figure 4.10   Electric monorail urban train, Sydney (photo by author).

used for nearly all train transport, except for those situations with the longest haul range versus traffic density.

4.1 A North American Reference Study

The scenario presented here is considered one to be realized around year 2060. It uses only renewable sources of energy. This is one option favored by many due to (a) its avoiding climate impacts from anthropogenic alteration of the greenhouse effect and (b) not having to deal with the unpredictable low-probability but high-consequence failures of large, centralized energy systems such as the nuclear ones. Many renewable energy systems are also ideal for illustrating the issues surrounding the handling of intermittency.

4.1.1 Contiguous United States

The reference model constructed considers the United States divided into a contiguous part plus detached parts such as Alaska and island communities (e.g., the Hawaii islands). It is based on two premises, one being the energy demand envisaged by 2060 and the other an inventory of renewable energy sources that may be exploited without conflict with sustainability or competing concerns (such as competing land use and other environmental impacts). However, these two premises do not fully specify the system, as there will typically be many ways of assembling an energy supply system by use of the given energy sources and in such a way that the expected demand can be satisfied at all times. Time simulation of the proposed systems is required, because both supply and demand vary with time. First, one may try to make the system work autonomously, and if this turns out to be impossible, one should specify which imports or joint operation schemes with neighboring systems are required to solve the problems.

Demands (here taken as delivered energy, as discussed previously) are conveniently divided according to energy quality (Sørensen 2010), from low-temperature heat to high-quality energy in the form of electricity or mechanical energy. Fuels have a quality depending on the temperature at which they are burned, as quantified by the second law of thermodynamics. Application of electricity or of mechanical or chemical energy does not entail any unavoidable losses, while going through heat, e.g., by combustion, at a given temperature T (measured in degrees Kelvin, K^{**}), leads to an unavoidable energy loss, L, given by

$$L = T_{ref}/T$$

where T_{ref} is the temperature of the surroundings (also expressed in K). If $T = T_{ref}$, no energy can be extracted, while, if T is much higher

---

* The temperature in Kelvin equals the temperature in degrees Celsius (centigrade) plus 273.15.

than T_{ref}, then the loss may approach zero. Of course, the specific technology employed may entail additional losses above those previously described as "unavoidable." Because fuels aimed for use in the transportation sector have to satisfy other specifications than stationary energy forms, they may be singled out as distinct from the electricity that can be used for electronics, for providing mechanical energy, for space cooling and refrigeration, and for delivering process heat at temperatures up to the highest ones. Several previous studies (see survey in Sørensen [2010]) used three ranges of heat usage, from the high temperatures (above 500°C) used for industrial processing of metals and minerals, to medium-temperature heat (100°C–500°C) used in light industries, to low-temperature heat (below 100°C) used for space heating and hot water at residential or commercial venues, and for process heat in some industries, e.g., for food processing. Because the renewable energy sources available in North America either produce electricity (wind, hydro, photovoltaic), are biofuels that may be combusted at any level of temperature, or are low-temperature heat sources (solar thermal collectors), the three temperature intervals used in earlier studies are not required, and the demand will here be specified in terms of just the following energy categories:

1. Low-temperature heat (for building or process use)
2. Electricity (including dedicated uses as well as cooling and production of all process heat above the low-temperature regime)
3. Gaseous fuels (such as hydrogen)
4. Liquid fuels (various biofuels)

Solid biofuels (such as firewood) are not considered because of the extreme air pollution caused by decentralized combustion. Use in high-temperature central power plants with pollution control equipment can reduce the emissions, but this is less interesting in a scenario where most of the energy sources (wind, hydro, photovoltaic) already provide electric energy that, if required, can be efficiently converted into low-temperature heat by use of heat pumps. Liquid biofuels also produce emissions undesirable to health and environment, but they allow better control even in decentralized applications, similar to the situation for the best current motor vehicles, which have equipment (catalysts, electrostatic filters) to reduce both sulfur and nitrous oxides and also particulate emissions. Once these vehicles use CO_2-neutral fuels, their remaining emissions may be considered tolerable, at least until true zero-emission vehicles (presumably fuel-cell–battery electric hybrids) become technically and economically viable (Sørensen 2012a). While short-rotation-time crops (such as cereals with straw residues) can be considered CO_2-neutral due to the short delay between assimilation and release, this is not the case for wood biomass and related residues, because here the assimilation of CO_2 may have occurred a century ago, when global warming was not an issue, as the climate system could still handle the level of emissions taking place.

The assumed US energy demand to be delivered in the 2060 reference scenario being simulated is shown in Figure 4.11. It is based upon the following assumptions (see Section 3.2.2. in Sørensen [2008a]): The 2060 average efficiencies along the energy-conversion chains between sources and end-use energy are assumed to equal those of the best commercially available technologies in 2005. This would generally imply three to five times less energy for the same service than the average in 2005. Many studies over the past decades have demonstrated that such high efficiency gains are not only technically realistic, but also economically viable (Beijdorff and Stuerzinger 1980; Sørensen 1982, 1991; Weizsäcker et al. 1997, 2009). Higher efficiency gain usually costs more money, but for each expected average energy price

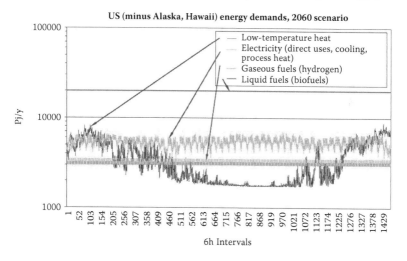

Figure 4.11   Continental US energy demand at the end user for low-temperature heat, electricity used directly or for cooling and process heat, and for gaseous and liquid fuels used in the transportation sector, according to 2060 reference scenario, as a function of time throughout a year and plotted by 6-hour intervals.

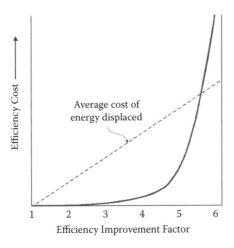

Figure 4.12 Typical behavior of the cost of improving the efficiency of a device converting or otherwise using energy to provide a desired service. The generic curve is based on references discussed in the text, with the efficiency base point (factor 1) corresponding to average equipment a few decades ago. Current levels of efficiency would be around a factor of 2, but with a large spread depending on the type of equipment and local positioning in terms of technological development.

level during the lifetime of the energy-converting equipment, there will be a range of efficiency measures up to some maximum that can be undertaken for a cost less than that of the energy they displace, as illustrated in Figure 4.12. The typical behavior is that the cost of efficiency improvements starts out at zero or a very low value, but at some point—after the easiest measures have been carried out—cost starts to rise quickly and at some point surpasses the value of the energy saved.

The reduced energy demand caused by efficiency improvements is countered by possible future higher activity levels and by introduction of novel energy-consuming technologies. For electricity use, this is taken to involve a doubling of demand by 2060, relative to 2005, thereby eating up about half of the efficiency improvement. For other energy demands, the activity increase is estimated to be more modest.

At the bottom of the low-temperature heat-demand time variation shown in Figure 4.11, a fairly constant level is describing hot water use in residential and commercial buildings, plus low-temperature process heat, e.g., that used in the food-processing industry. Increased demands for space heating occur during winter, estimated using temperature data from ECMWF (2013) and an energy expenditure of 24 W/cap. per degree C, the temperature has to be raised (Sørensen, 2010; see also discussion in Chapter 9.1 below). The same model is used in all the following scenarios. The electricity need similarly comes from (a) residential equipment (washing machines, refrigerators, entertainment electronics and computers, plus battery charging for several types of appliances and devices) and (b) commerce and industry, using many types of office equipment similar to that used in households, plus the electricity needed to drive motors and produce medium- and high-temperature process heat, plus a range of assembly-line equipment powered by electricity. Mechanical energy and process heat for raw-materials extraction and recycling are assumed to be provided by transformation of electric energy. Finally, the energy used for vehicles serving to satisfy transportation needs is in this renewable energy scenario, divided between gaseous fuels such as hydrogen (for fuel-cell–battery hybrid cars) and liquid fuels such as ethanol and biodiesel. The ratio between gaseous and liquid fuels is a fairly arbitrary guess as to the penetration of various new mobile propulsion systems by 2060, and the fairly low share of hydrogen for the United States is chosen due to the priority given to using renewable electricity from wind, solar, and hydro sources for dedicated stationary uses, leaving fairly little to be converted into hydrogen. At the same time, since industrial process heat is also considered to be covered by electricity, there are abundant biofuels for use in the transportation sector. Of course, these preferences could be altered. The totals corresponding to the time distributions shown in Figure 4.11 are given in Table 4.1.

The electricity energy sources available for the US 2060 scenario are (a) onshore and, at coastal locations, offshore wind power from turbines restricted here to a foundation depth of less than 50 m (disregarding the possibility that future wind utilization from floating platforms may become possible[*]), (b) solar photovoltaic panels (both building-integrated and collector fields placed on marginal land), and (c) hydropower plants. For onshore wind, the data summarized in Chapter 3 are used, based on 6-hour time series provided by general circulation models as part of the "blended" data set constructed by NCAR (National Center for Atmospheric Research 2006; Sørensen 2008c), folded with

---

[*] The actual calculations make use of a geographical grid with cells of 0.5° × 0.5° longitude–latitude extent, and only cells with fractions of both land and sea are included in the offshore wind estimate. The implied limit to the distance of wind installations from shore is 56 km, and the average distance is 20–25 km.

Table 4.1   Summary of 2060 Reference Scenario for Contiguous United States (PJ/y)

	LOW-T HEAT	ELECTRICITY	GASEOUS FUELS	LIQUID FUELS
Delivered energy demand	3435	5423	3126	20,000
Onshore wind-power production		4470		
Offshore wind-power production		2678		
Hydropower production		849		
Photovoltaic power production		2789		
Biofuels from agricultural residues				7686
Biofuels from forestry residues				10,271
Biofuels from aquaculture				2046
Solar thermal energy produced	3105			
Electricity for dedicated uses		5373		
Electricity for hydrogen production	84	3735	3362	
Electricity to heat by heat pumps	1509	503		
Liquid biofuels for transportation				20,000
Hydrogen for use in vehicles			3126	
Solar thermal heat used directly	1355			
Low-temperature heat from stores	487			
Discarded or lost solar heat	1263			
Potential export		1175		

Note: The assumed 2060 population is 347.543 million.

the wind-turbine power-conversion characteristics. There are variations at timescales below 6 hours, but feeding power from a large number of turbines into a national grid will smooth out such fluctuations (see the power duration curves for one and several wind turbines shown as Figure 5.6 in Sørensen et al. [2001]). The number of sites exploited for wind harvesting is fairly modest, taken as a rotor-swept area constituting 0.02% of the horizontal land area of the United States.[*] In terms of density of wind turbines, this roughly corresponds to the number of wind turbines currently operating on land in Denmark (although many of these are decades old and smaller in size than the ones envisaged in the 2060 model). The visual impact of wind turbines increases less than proportionally to their size (see Chapter 10 in Sørensen [2012b]).

The electric power production from offshore wind farms is also derived by folding the 6-hour time series of the NCAR blended data

---

[*] In an earlier survey of energy resources for the USA (Chapter 6 in Sørensen, 2010), it was found that a wind-swept area of 0.01% of the land area was insufficient for covering expected demands.

set, which over sea is primarily based on scatterometer (a sweeping microwave radar device) satellite observations (Sørensen 2008a, 2008c), with the power curve of contemporary, multimegawatt-size wind turbines, after having scaled the scatterometer data (pertaining to a height of about 3 m) to a typical wind turbine hub height of 70 m (by the method described in Sørensen [2008b]). The wind turbine towers are assumed to stand on a foundation reaching the sea floor, for which reason the mentioned maximum depth is imposed. However, the windswept area is restricted to 0.1% of the relevant ocean area due to consideration of competing uses (such as sailing routes, fishery, military).

Solar power in the 2060 US scenario is derived from building-integrated (on rooftop or facade) solar cells and some centralized solar fields. However, the source inflow for the building-integrated systems is taken to comprise solar radiation on just 2 m^2 per capita, out of an average availability of about 60 m^2 per capita, of which some 25% would be suitably oriented (more or less toward the south) rooftop or facade areas with modest shading from other buildings or structures. The solar collectors are assumed to be of the PVT type (photovoltaic and thermal combined, see Sørensen [2000, 2001]), where the waste heat from 10%–20% efficient solar cells is partially collected by a backside panel or by an airflow channel below a top glass cover (Figure 4.6) and transferred as in a thermal collector, operated in connection with a thermal storage unit, here assumed to be of modest size (such as a large hot-water tank) (see Sørensen [2010, chap. 6]). This holds only for the building-integrated part, because separate solar farms, assumed to be installed on 0.1% of all marginal land in the United States (the total marginal land mass being 1.4 million km^2) (US Geological Survey 1997), would require district-heating lines to heat load areas in order to exploit the PVT coproduction.

This is not required or desirable in the US model, as the low-temperature heat demand is modest in 2060, due to assumed building-efficiency improvements. The same is a contributing reason for hot-water stores in buildings to be kept at low volume. Seasonal heat storage would require communal stores, because the ratios of storage surface area (heat loss) to storage volume (amount stored) are generally unfavorable for small, single building installations. In any case, small storage volumes will cause summer temperatures in the store to be high (making collection efficiency low), while large storage volumes will cause winter temperatures in the store to be lower than that of the heat demanded for space heating or hot water. As a result of these conditions, the solar heat derived by the PVT systems of the 2060 US scenario is only a little larger than the photovoltaic power production, far below the theoretical maximum of two to five times the electric output, especially during summer. To calculate the incoming solar radiation on the PVT panels, instead of using actual time series (as for wind), a very simplified model is employed, based on cyclical variations of the level of solar radiation over each day, taking into account the geographical and seasonal variations in the duration (hours) of daylight and the angle of incidence, and then combining this with a stochastic noise multiplier representing cloud cover and air turbidity. This multiplier is random except for a 20% correlation with the value at the preceding time step (6 hours earlier). One reason for using this simple model is that solar radiation data on inclined surfaces is available only at selected locations, and that the algebraic model approach for inclined surfaces outlined in Sørensen (2010) would appear to be nearly as crude as the random model. This is further discussed in the scenarios presented in Chapter 13.

As regards hydropower, it is assumed that, for environmental reasons, no new dams will have been established in addition to those already existing. Although small-scale hydropower could be somewhat expanded, the assumed hydropower production is taken at the present level. The energy simulation model does not consider variations in water inflow to hydro reservoirs, but simply assumes a constant production over the year due to the smallness of the hydro contribution relative to that of wind and solar cells. Had this not been so, the hydropower could be used to smooth variations in wind and solar source flows, but in the US 2060 scenario, this has not been found necessary, as the dedicated electricity demands are found at all times to be satisfied directly. The reason for this is that the fluctuations in wind and solar resources are absorbed by the substantial additional amounts used to produce hydrogen, which do not have to be synchronous with demand due to the assumed availability of hydrogen stores.

For satisfying the demands of the transportation sector, hydrogen for fuel-cell operation appears to be the most environmentally desirable solution, if current technical problems of fuel-cell lifetime and performance can be solved (Sørensen 2012a). The 2060 reference scenario for the contiguous United States cautiously assumes a fairly low penetration of hydrogen-fueled vehicles by 2060. One motivation for this is that although hydrogen can be produced on the basis of biomass, the alternative of using most biomass for liquid biofuels seems more attractive, because it does not require both an expensive fuel and an expensive fuel cell to use it, but will allow combustion in an ordinary car engine. This argument depends on the ability to curb emissions from biofueled engines. The alternatives to biofuels are the use of batteries (expensive) connected to electric motors, or the production of hydrogen by electrolysis (technically achieved by reverse operation of some kind of fuel cell, currently mostly alkali types, also an expensive, but fully established technology with high conversion efficiency). The systemic expense can be reduced if the same fuel cell is operated in both directions, i.e., also to produce electricity from hydrogen. A decentralized scenario with this type of fuel-cell use (by stationary units in buildings) was proposed in the Danish Hydrogen Program (Sørensen et al. 2001), but the need for this seems less urgent in the US 2060 reference scenario, because it does not exhibit a shortage of electricity for satisfying time-urgent demands. In all cases, conversion of hydrogen back to electricity is what takes place in fuel-cell vehicles, where the actual shaft power is delivered by an electric motor.

Thus most of the energy for transportation in the reference 2060 US scenario is delivered by biofuels, derived from agricultural residues, forestry residues, and aquaculture at conversion efficiencies assumed to be 45%–51% (highest for wood waste to methanol, lower for ethanol or biodiesel production). Because these fuels can be stored, only the annual total production is estimated and compared to total demand. The total biomass production is calculated by use of a model developed by the Woods Hole group (Melillo et al. 1998). The version used here does not consider artificial irrigation, which may have undesirable long-term effects on the water cycle, although some level of irrigation may be performed sustainably (Sørensen 2010). The subsequent model for the distribution of potential net biomass production on land areas classified as farmland, forests, etc., and making use of the options for conversion to food, fuels, and industrial raw materials, with possibilities for recycling, is taken from Sørensen (2010, chap. 6). It is important to notice that many biomass usage avenues employed today can also be maintained with biofuel production, because the residues from the biofuel conversion can still be returned to the agricultural fields or stables (like straw) or made available to industries for use as fibers and special components (e.g., for glue manufacture). Only the current combustion of straw and forestry wood fuel will no longer be possible, as this practice will be replaced by the more valuable production of liquid fuels for transportation. From a sustainability point of view, returning nutrients to the fields is a necessary requirement.

For the 2060 US model, it is assumed that residues corresponding to 50% of the biomass production in agriculture are collected and converted into biofuels, and that residues corresponding to a third of the forestry biomass production are collected and converted into biofuels (e.g., not touching national parks and national forests, although energy use of their residues could probably be accepted). To reach the 20,000 PJ/y assumed to be demanded, an additional 2046 PJ/y must be derived from aquaculture, which takes 87% of this kind of resources, as identified by the previously mentioned model, using a crude estimate where only those $0.5° \times 0.5°$ grid cells of the geographical model that contain both land and waterways/ocean are considered available for biomass farming, implying a distance of 20–25 km from shore. The productivity per area estimate for the land fraction is considered achievable also for ocean farming on the remaining area, with suitable crop choice and fertilization. It is not unlikely that by year 2060, ocean farming of kelp and other seaweeds for food will have been globally adopted.

While, as mentioned, no detailed modeling of biofuel and solar thermal stores is considered necessary here, the storage in hydro reservoirs and hydrogen stores is part of the simulation model. For each run, the magnitude of the store (i.e., its maximum energy holding) is adjusted until the level at the end of the simulated year is the same as the one at the simulation start. For the present simulation, with no variations in hydrogen demand, a total hydrogen storage capacity of just over a day of production suffices, but with realistic (including interannual) demand variations, a couple of weeks' worth of storage would be better. Due to the relative stability of wind-power production, no hydro storage capacity is required in the simulation, although some is actually already present in the United States. This implies that a surplus of electricity is produced that, in the model, cannot be stored due to the minimal assumptions on hydrogen or elevated water stores available, and which therefore becomes available for export to neighboring countries or regions. If the hydro or hydrogen stores were made larger, their filling at the end of the year would be higher than at the beginning, which of course in the hydrogen case would open for alternative scenarios with more transportation energy covered by hydrogen and less by biofuels, allowing those to be exported or left unused.

The annual total availability of renewable energy for the 2060 scenario is summarized in Table 4.1, split on individual sources and energy quality. Table 4.1 also gives annual totals for the major energy conversions needed to make supply meet demand, notably conversion of electricity to hydrogen or to low-temperature heat via heat pumps. The heat pump coefficient of performance (COP) is conservatively set at three, although many heat pumps already today are capable of delivering four to five times the electricity input as heat. The scenario heat pump systems are only operated intermittently and rely on air (e.g., from building ventilation systems) or heat from the solar stores as their cold reservoir, rather than the more expensive underground (at frost-free depth) pipelines considered to give optimum performance. The electric power export potential is 1175 PJ/y.

Figure 4.13 shows the 6-hour time evolvement of the various uses to which produced electricity is assigned. Because the power from wind

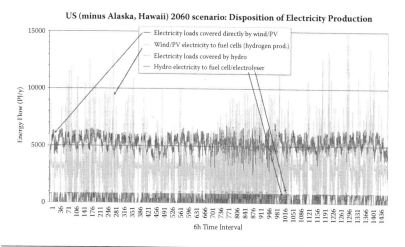

Figure 4.13   Continental US disposition of electricity produced by wind power, photovoltaic panels, or hydro energy for direct uses and for hydrogen production, according to 2060 reference scenario, as a function of time throughout a year and plotted by 6-hour intervals.

and photovoltaic systems is uncontrollable, it is given first priority for satisfying dedicated power loads. This is seen to give full coverage of the loads except for the late summer period and a few scattered hours during the remaining part of the year. These demands are covered by hydro energy, which, directly or through the elevated water reservoirs, is always capable of delivering the electricity deficit, assuming the availability of the necessary power transmission capacity. The hydro systems further deliver some power for hydrogen production intermittently throughout the year. The fourth data series in Figure 4.13 is showing surplus electricity from wind or photovoltaic generators being used for hydrogen production (assumed to take place by reversed fuel-cell operation). An intermittent, but sizable, amount of hydrogen is produced during spring and fall, mostly by wind, while the surplus during the rest of the year is more modest but still substantial. The net surplus power production available for export is shown in Figure 4.14.

The time development of the filling level in the hydrogen stores (constantly at their assumed maximum capacity) is shown in Figure 4.15, along with the demands for hydrogen and liquid fuels by the transportation sector. The time variations of these loads are considered unimportant, and only an hour-of-the-day variation in filling automobile tanks is indicated. Figure 4.16 shows the time-series of heat disposition. The heat production from the thermal part of solar PVT systems can be quite substantial even during winter, although not always available. This is due to the assumption that there

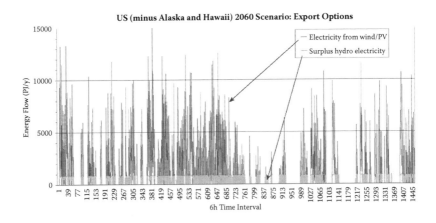

Figure 4.14   Potential energy export options available to the continental United States, according to 2060 reference scenario, as a function of time throughout a year and plotted by 6-hour intervals.

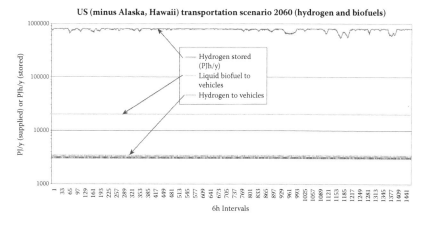

Figure 4.15   Coverage of continental US transportation needs by liquid biofuels or by hydrogen, along with storage level of established hydrogen storage facilities (primarily placed underground), according to 2060 reference scenario, as a function of time throughout a year and plotted by 6-hour intervals. The energy sources that are filling the hydrogen stores are depicted in Figure 4.13.

are more PVT panels in the south of the contiguous United States (taking radiation data from an average latitude of 30° N) than in the north, due to the higher electricity production and winters with many clear days. Combined with low solar-heat-store temperatures and the ensuing high efficiency of solar-to-heat conversion (either the water

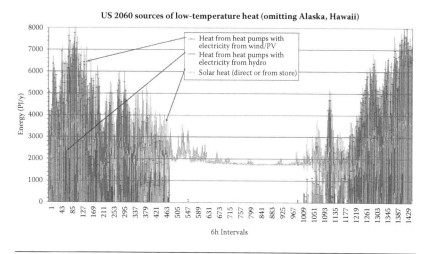

Figure 4.16   Coverage of continental US demands for low-temperature heat, by solar thermal installations (directly or through heat stores), or by heat from wind/PV and hydro electricity administered to heat pumps (with an assumed ratio of 3 between heat output and electricity input), according to 2060 reference scenario, as a function of time throughout a year and plotted by 6-hour intervals.

of the stores is circulated through the collector panels, or the collector flow is by air exchanging heat with the heat-store medium), this gives a substantial thermal production that may not be useable at the southern latitudes. Throughout the summer period, the solar thermal system is more than capable of satisfying heat demands at all times and at all latitudes. (During summer, there is little or no space-heating demand but still a substantial demand for hot water in buildings and industry.) The autumn situation is not symmetrical with that of spring, because the heat-store temperatures are now higher, depending, as mentioned, on the storage size. The issue has been clarified by simulation of a number of systems with different sizes of heat stores, as described by Sørensen (2010). The deficit in required heat coverage (mostly space heating in northern United States) is made up for by application of electricity through heat pumps. Figure 4.16 shows the heat derived from heat pumps employing wind-, PV-, or hydropower, while Figure 4.13 showed the electricity inputs to the heat pumps.

The fairly modest electricity export potentials shown in Figure 4.14 occur intermittently throughout the year and are quite sizable during early summer, where the PV power production is largest. An alternative to export would be to produce more hydrogen in the United States and use it in the transportation sector or export it through pipelines, but the reference scenario considers power transmission over long distances to be less expensive than hydrogen pipeline transmission. This is not a firm statement, as the seasonality of the US surplus electricity will probably force the receiving buyer to establish large hydrogen reservoirs or other stores in order to cover a demand that likely is more or less uniformly distributed over the year. The timewise supply–demand mismatch has to be addressed anyway, either in the exporting or in the importing country, depending on the cost structure and availability of storage options. The latter could be different in the selling and buying country, e.g., regarding the availability of underground hydrogen storage sites, which are much less expensive when suitable aquifers or salt domes are present compared to storage sites requiring excavation of rock (Sørensen 2012a).

4.1.2 High-Demand Alternative Scenario for the Contiguous United States

The assumptions regarding demands in the reference scenario may not match the ones actually evolving or being politically selected, so alternatives should be explored to satisfy the scenario method's prescription of presenting a range of scenarios for democratic debate (Sørensen 2010). In the US reference case, considerable emphasis has been put on investments in energy efficiency before assembling the energy production system, as this was argued (Figure 4.12) to constitute the lowest-cost solution. However, the technical conversion system is only part of the scenario, another one being the future development of energy-consuming activities. As regards transportation, it is difficult to envisage much higher activity than the one assumed in the scenarios, as that would imply that citizens would have to spend a lot more time in cars on the roads than they already do, and probably even more than the increase in transportation activity, because road construction and traffic management will hardly be able to accommodate greatly increased activity without more congestion and delays. For low-temperature heat, a higher consumption level could be envisaged if buildings were not being better insulated in the future, despite the clear economic incentive. A well-constructed building with proper materials and architecture, as well as high levels of insulation and controlled air exchange, actually not only reduces heat demand in winter, but also reduces cooling needs during summer, making the investment doubly beneficial. It is therefore hard (although far from impossible) to envisage scenarios with substantially higher heat demands than those of the reference scenario. That brings us to the last category, electricity demand. It is here that the scenario uncertainty is largest, as one could envisage new technologies coming into play that are not, as the scenario assumes, continuations of current trends (e.g., in microelectronics), but are radically novel and with a functioning not anticipated at present. The appearance of such "unknown" demands is worth an alternative scenario, and although being unknown, it seems a good guess that the extra demand will be in the form of electricity or some equivalent energy kind.

In consequence, a scenario simulation of the 2060 contiguous United States was run with electricity demands doubled in order to see the implications for the success of using only renewable energy sources. This type of investigation would come under the heading of exploring "the system's resilience to altered assumptions," a concept introduced by practitioners of system life-cycle analysis (Sørensen

2011), and some observers would even consider the high-demand scenario as quite likely to evolve.

Table 4.2 summarizes the outcome at the same level of aggregation as Table 4.1. A good deal of the additional electricity supply can be satisfied by wind, hydro, and photovoltaic power, but there remains a deficit that, in the scenario, is covered by a slight increase in the use of biofuels, achieved by raising the fraction of biomass collected from agriculture from 50.0% to 54.5%. As a further consequence of the higher direct use, there is less electricity that may be converted to hydrogen or used in heat pumps for low-temperature heat. Most of the missing heat from heat pumps is compensated by making use of the associated heat from the additional biofuels combusted in combined heat and power plants (CHP). Actually the additional power and heat delivered by CHP could probably be produced with fewer resources

Table 4.2 Summary of 2060 Reference Scenario for Contiguous United States with Twice the Electricity Demand (PJ/y)

	LOW-T HEAT	ELECTRICITY	GASEOUS FUELS	LIQUID FUELS
Delivered energy demand	3435	10845	3126	20,000
Onshore wind-power production		4470		
Offshore wind-power production		2678		
Hydropower production		849		
Photovoltaic power production		3175		
Biofuels from agricultural residues				9230
Biofuels from forestry residues				12,670
Biofuels from aquaculture				2046
Solar thermal energy produced	3515			
Electricity for dedicated uses		9402		
Electricity for hydrogen production		1299	1169	
Electricity to heat by heat pumps	63	21		
Liquid biofuels for transportation				20,000
Liquid biofuels for CHP or furnace	1298	1443		3946
Hydrogen for use in vehicles			1052	
Solar thermal heat used directly	1529			
Low-temperature heat from stores	545			
Discarded or lost solar heat	1168			
Potential export		451		
Import required			2074	

Note: Other assumptions unchanged except 9% more agricultural residues used.

if biomass residues were combusted directly in the CHP plants rather than first being converted to biofuels, which is only necessary for the part to be used in vehicles. The available associated heat from fuel-cell conversion is small and assumed to be unused because of the distance between the point of production and the heat loads. Associated heat from combined power and heat plants, in contrast, is large and warrants establishment of district heating lines to major heat-demand areas (such as cities). Because the hydrogen that can be produced with the electricity available in this scenario is insufficient to cover even the 20% of transportation energy demand assumed for the contiguous United States, there remains an import need amounting to 66% of the gaseous fuel demand. Clearly, the deficit could also have been listed in the form of electricity, in case the missing hydrogen is supplied from imported electric power. The alternative of further increasing the use of domestic biomass for producing vehicle fuels begins to look questionable, considering its need to collect most of the biomass not eaten (including residues at the end user) in a sustainable way.

This scenario variant has shown how delicate the balancing between system components is and supports the claim that performing scenario simulations is highly beneficial for the construction and understanding of optimally functioning future energy systems.

4.1.3 Canada

The Canadian energy demand–supply system in 2060 is simulated with basically the same methodology as that for the United States. However, the variant with doubled electricity demand and with reduced wind-power utilization (swept area 0.01% of land surface area, mainly in the southern part) is here considered the most likely one to emerge because (a) electricity is abundant due to the wealth of renewable energy sources relative to demand in Canada and (b) because exploiting wind resources in the far north has been reduced for both practical and environmental reasons. Tables 4.3 and 4.4 show the summary results of the reference scenario assumptions and those of the doubled demand and halved onshore wind power, respectively. Figures 4.17–4.20 give the energy demands, the disposition of electricity and heat production, and the export potentials for the scenario with doubled power demand and reduced onshore wind production,

Table 4.3 Summary of 2060 Reference Scenario for Canada (PJ/y)

	LOW-T HEAT	ELECTRICITY	GASEOUS FUELS	LIQUID FUELS
Delivered energy demand	515	550	833	1000
Onshore wind-power production		5985		
Offshore wind-power production		4415		
Hydropower production		1170		
Photovoltaic power production		309		
Biofuels from agricultural residues				945
Biofuels from forestry residues				5611
Biofuels from aquaculture				900
Solar thermal energy produced	346			
Electricity for dedicated uses		550		
Electricity for hydrogen production		834		
Electricity to heat by heat pumps	309	103		
Liquid biofuels for transportation				1000
Hydrogen for use in vehicles			833	
Solar thermal heat used directly	192			
Low-temperature heat from stores	15			
Discarded or lost solar heat	139			
Potential export		9786	0	2278

Note: The assumed 2060 population is 36.352 million.

respectively. For biomass resources, the added transport cost of moving biofuels produced at high latitudes to demand sites has been taken into account by assuming that only a third (corresponding to production in southern Canada) of the agricultural and aquaculture biomass resources estimated by the Woods Hole model without irrigation is used for biofuel production, and that only a quarter of the forestry biomass residues (or growth, assuming no annual change in standing crop) is converted into biofuels. Decentralized solar PVT systems are again taken as 2 m² per capita, while the contribution of central PV plants on marginal land is assumed to be only 0.05% of the marginal areas, once more due to the practical difficulty in using areas that are far to the north. Electricity surpluses that may be used for hydrogen production are generously available; thus the transportation demand is divided into two equal parts, to be satisfied by liquid and gaseous fuels. The differences between the energy needs for vehicles (given in Tables 4.3 and 4.4) are due to the different efficiencies of biofuel vehicles and hydrogen vehicles, which could alternatively be

Table 4.4 Summary of 2060 Reference Scenario for Canada with Twice the Reference Electricity Demand and about Half the Onshore Wind Production (PJ/y)

	LOW-T HEAT	ELECTRICITY	GASEOUS FUELS	LIQUID FUELS
Delivered energy demand	515	1101	833	1000
Onshore wind-power production		2992		
Offshore wind-power production		4415		
Hydropower production		1170		
Photovoltaic power production		309		
Biofuels from agricultural residues				945
Biofuels from forestry residues				5611
Biofuels from aquaculture				900
Solar thermal energy produced	346			
Electricity for dedicated uses		1101		
Electricity for hydrogen production		834		
Electricity to heat by heat pumps	288	96		
Liquid biofuels for transportation				1000
Hydrogen for use in vehicles			833	
Solar thermal heat used directly	192			
Low-temperature heat from stores	36			
Discarded or lost solar heat	118			
Potential export		6250	0	2278

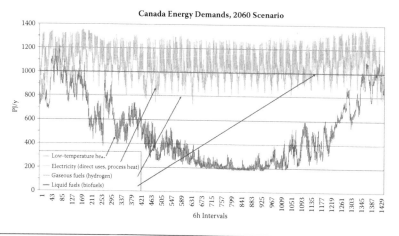

Figure 4.17 Canadian energy demand at the end user for low-temperature heat, electricity used directly or for cooling and process heat, and for gaseous and liquid fuels used in the transportation sector, according to the 2060 scenario with doubled electricity demand relative to the reference value, as a function of time throughout a year and plotted by 6-hour intervals.

hydrogen–battery hybrids, with some electricity charging the batteries rather than producing hydrogen. This remark pertains to hydrogen vehicles in all the countries modeled.

As Figure 4.17 shows, the average difference between winter and summer heat use is, as expected, larger in Canada than in the contiguous United States. Fuel demands for the transportation sector have, as mentioned previously, been divided equally between gaseous fuels (hydrogen) and liquid fuels (biofuels). As in the US case, the split between gaseous and liquid fuels is largely an arbitrary choice. Figure 4.18 is surprise free, with renewable-energy satisfaction of both direct electricity needs and hydrogen production (into hydrogen stores), even without use of the available hydro reservoir stores. The reason is that the power surplus is so large that wind (and a little photovoltaic) power, even after the halving of the installed capacity, can easily cover all domestic demand, so that in this schematic scenario, all potential hydropower production is available for export, along with a large but fluctuating part of the wind power, as shown in Figure 4.20. Of course, in this situation there is no need to distinguish between the sources of power. It is just a feature of the model used to give first priority to the variable wind power and second priority to the partially controllable hydropower. Also for biofuels, Canada is seen

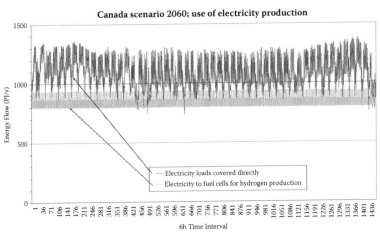

Figure 4.18 Canadian disposition of electricity produced by wind power, photovoltaic panels, or hydro energy for direct uses and for hydrogen production, according to the 2060 scenario with doubled electricity demand and onshore windswept area equal to 0.01% of land surface, as a function of time throughout a year and plotted by 6-hour intervals.

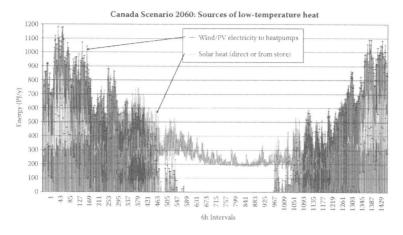

Figure 4.19 Coverage of Canadian demands for low-temperature heat by solar thermal installations (directly or through heat stores) or by heat from heat pumps (the heat delivery assumed to
be 3 times the wind/PV electricity input), according to the 2060 scenario with doubled electricity
demands and halved onshore wind production, as a function of time throughout a year and plotted
by 6-hour intervals.

in Figure 4.20 to have an export surplus. Figure 4.19 shows that solar
thermal production can satisfy heat demands in summer, but that heat
from heat pumps is required from October to May, with production
peaking during the winter months, with modest assistance derived
from the solar panels. Solar heat production fluctuates greatly in winter as well as during the spring and autumn months.

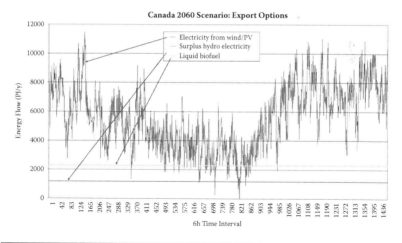

Figure 4.20 Potential energy export options available to Canada, according to the 2060 scenario
with doubled electricity demand and halved onshore wind production, as a function of time throughout a year and plotted by 6-hour intervals.

The large export potential available in the Canadian energy system is explored further in Chapter 5, looking at the advantages that may be derived from international trade of energy among the North American countries.

4.1.4 Alaska and Greenland

The Alaska and Greenland 2060 scenario simulations proceed as for the contiguous United States and Canada, except that availability of solar energy and biomass residues diminishes with latitude. For Greenland, biomass resources are taken as essentially zero, and solar energy is so seasonal that it was deselected in a scenario with plenty of alternative options for the tiny population of Greenland. For Alaska, there is some solar potential in the south, and forestry resources are quite large, but agricultural resources are virtually absent. As in Canada, only 25% of the forestry residues are considered for energy. The simulation results presented in Tables 4.5 and 4.6 both assume the scenarios with doubled

Table 4.5   Summary of 2060 Reference Scenario for Alaska with Twice the Reference Electricity Demand and about Half the Onshore Wind Production (PJ/y)

	LOW-T HEAT	ELECTRICITY	GASEOUS FUELS	LIQUID FUELS
Delivered energy demand	9.2	23.6	14.6	21.0
Onshore wind-power production		615.2		
Offshore wind-power production		963.8		
Hydropower production		151.5		
Photovoltaic power production		74.9		
Biofuels from agricultural residues				0.5
Biofuels from forestry residues				2218.9
Biofuels from aquaculture				379.0
Solar thermal energy produced	51.3			
Electricity for dedicated uses		23.6		
Electricity for hydrogen production		14.6		
Electricity to heat by heat pumps	0.3	0.1		
Liquid biofuels for transportation				21.0
Hydrogen for use in vehicles			14.6	
Solar thermal heat used directly	4.6			
Low-temperature heat from stores	4.3			
Discarded or lost solar heat	42.4			
Potential export	0	1765.5	0	1088.7

Note: The assumed 2060 population is 0.78 million.

Table 4.6   Summary of 2060 Reference Scenario for Greenland with Twice the Reference Electricity Demand and about Half the Onshore Wind Production (PJ/y)

	LOW-T HEAT	ELECTRICITY	GASEOUS FUELS	LIQUID FUELS
Delivered energy demand	1.5	2.2	1.4	1.9
Onshore wind-power production		651.4		
Offshore wind-power production		342.4		
Hydropower production		71.3		
Photovoltaic power production		(20.7)		
Biofuels				0
Solar thermal energy produced [a]	(12.5)			
Electricity for dedicated uses		2.2		
Electricity for hydrogen production		1.4		
Electricity to heat by heat pumps	1.5	0.5		
Liquid biofuels for transportation				0
Hydrogen for use in vehicles			1.4	
Import requirement				1.9
Potential export	0	1060.8	0	0

Note: The assumed 2060 population is 72,000.
[a] The potential solar energy production in parentheses is not used in the scenario.

electricity demand and halved onshore wind, relative to the global reference methodology from Sørensen (2008a,b). The thinking behind this assumption is that people at higher latitudes probably spend more time indoors, using more lights and electric equipment.

It is seen from Table 4.5 that most of the solar heat produced in the Alaska scenario cannot be used at the time of production, but has to stay in heat stores for some time. Still, the remaining heat to be covered by electricity through heat pumps is modest. Electricity from wind and hydro are huge compared to the demands of the Alaska 2060 population, and so is biofuel production from forest residues. This entails a large potential for export. In Greenland, there is solar radiation available during the long summer days, but the scenario chooses to use heat pumps to supply low-temperature heat, due to the more evenly varying seasonality of the wind and hydro sources already used for producing electricity. Because the wind-power production is some 300 times the demand, the known seasonality of the Greenland hydro production was not investigated. As indicated in Table 4.6, there will be large amounts of electricity available for export (if energy transport or transmission routes can be established).

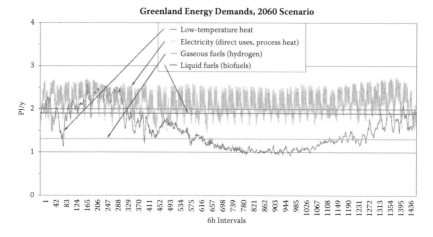

Figure 4.21   Greenland energy demand at the end user for low-temperature heat, electricity used directly or for cooling and process heat, and for gaseous and liquid fuels used in the transportation sector, according to the 2060 scenario with doubled electricity demand relative to the reference value, as a function of time throughout a year and plotted by 6-hour intervals.

Figure 4.21 shows the Greenland energy demand variation dominated by high space-heating needs and, over most hours of the year, the even higher electric power demands. The doubling of the values obtained with the reference assumptions of Sørensen (2008a) can, in addition to the indoor occupancy argument given here, be justified by a possible 2060 location of advanced mineral resource industries in Greenland, in line with current political ambitions for a future Greenland economy independent of subsidies from Denmark. Figure 4.22 shows the coverage of electricity demands, making use of the abundant wind-power resources and using hydrogen in the transportation sector. Smoothing of the uneven wind-power production by use of hydrogen stores or hydro reservoirs is required only for the export share, due to savings in transmission cost. The 2060 low-temperature demands are, as shown in Figure 4.23, entirely covered by heat pumps, considered to constitute the cheapest alternative, and thus disregarding potential summer production of heat by solar collectors. Figure 4.24 shows that Greenland by 2060 may have a very substantial electricity export potential, but that the liquid fuels assumed to cover half of the transportation demands will have to be imported (if they have to be of renewable origin and not based on the rather sizable Greenland oil resources), because the biomass production in Greenland is minute.

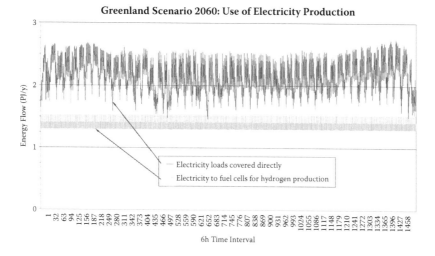

Figure 4.22   Greenland disposition of electricity produced by wind power, photovoltaic panels, or hydro energy for direct uses and for hydrogen production, according to the 2060 scenario with doubled electricity demand and onshore windswept area equal to 0.01% of land surface, as a function of time throughout a year and plotted by 6-hour intervals.

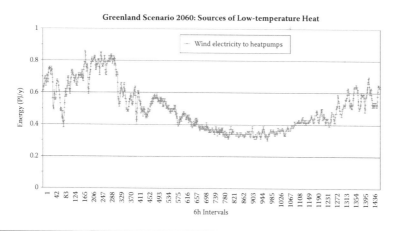

Figure 4.23   Coverage of Greenland demands for low-temperature heat by electricity administered to heat pumps (the heat delivery from which is 3–5 times the electricity input shown), according to the 2060 scenario with doubled electricity demands and halved onshore wind production, as a function of time throughout a year and plotted by 6-hour intervals.

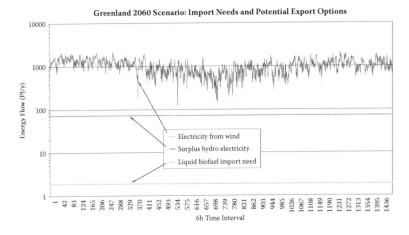

Figure 4.24   Energy import requirements (biofuels) and potential energy export options (electricity) available to Greenland, according to the 2060 scenario with doubled electricity demand and halved onshore wind production, as a function of time throughout a year and plotted by 6-hour intervals.

4.1.5 Mexico

The reference scenario for Mexico assumes lifestyles and thus per capita demands to be the same as, for example, in Europe, optimistically foreseeing a future where problems contributing to slow development in certain areas (notably in the northeastern part of Mexico hosting militant mafia organizations funded by the sale of drugs to the United States, or in the southwestern part with repeated incidence of violent insurgence) have been dealt with, a possibility perhaps encouraged by current trends in a few Mexican regions, such as the (offshore) oil-rich Yucatan peninsula. The assumed demands are shown in Figure 4.25 and Table 4.7. The renewable energy supplies indicated in Table 4.7 are dominated by direct solar energy and an amount of biomass that is large enough to cover the transportation needs, for which reason the demand for hydrogen fuel has been set at zero. However, this does not mean that hydrogen is absent from the Mexican energy system, because it can offer use as a storage medium, noting that the abundant electricity production, primarily from photovoltaic panels, is sometimes zero (e.g., at night) while electricity demand is not, requiring some medium capable of storing electric power, releasing power, and then later regaining it, albeit with losses.

The energy production by wind takes a swept area equal to 0.02% of the total land area plus, as in the previous cases, 0.1% of the

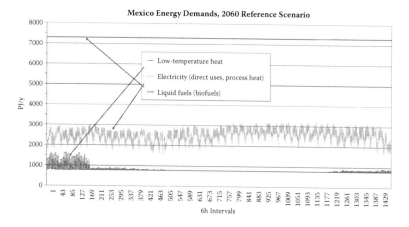

Figure 4.25 Mexico energy demand at the end user for low-temperature heat, electricity used directly or for cooling and process heat, and for liquid fuels demanded in the transportation sector, according to the 2060 reference scenario, as a function of time throughout a year and plotted by 6-hour intervals.

Table 4.7 Summary of 2060 Reference Scenario for Mexico (PJ/y)

	LOW-T HEAT	ELECTRICITY	GASEOUS FUELS	LIQUID FUELS
Delivered energy demand	805	2544	0	7301
Onshore wind-power production		846		
Offshore wind-power production		1169		
Hydropower production		285		
Photovoltaic power production		2563		
Biofuels from agricultural residues				1935
Biofuels from forestry residues				5101
Biofuels from aquaculture				720
Solar thermal energy produced	4777			
Electricity for dedicated uses		2414		
Electricity for hydrogen production		235		
Liquid biofuels for transportation				7301
Hydrogen used in fuel cells [a]	(65)	130	217	
Heat directly from solar panels	473			
Heat from local thermal stores	332			
Discarded or lost solar heat	3972			
Potential export		2214	0	0

Note: The assumed 2060 population is 154.12 million.

[a] Most associated heat from fuel cells is not made useful, as it appears at times where demand is already met, and often far from the locations where heat is demanded. An alternative to introducing hydrogen for coping with intermittent power production is to use some additional biofuels in power plants or CPH plants.

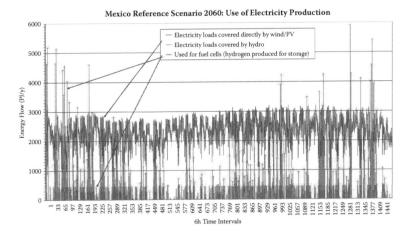

Figure 4.26   Mexican disposition of electricity produced by wind power, photovoltaic panels, or hydro energy for direct uses and for hydrogen production, according to the 2060 reference scenario, as a function of time throughout a year and plotted by 6-hour intervals. The onshore windswept area equals 0.02% of the land surface, and an assumed 8 m² of decentralized solar PVT collectors are installed per capita, plus central photovoltaic plants on 0.2% of the marginal land area in Mexico.

offshore areas near the coast. Because solar energy in Mexico is the most abundant renewable energy form and has less seasonal variation than in the higher-latitude North American countries considered in previous discussions, as much as 8 m² per capita, building-integrated solar PVT collectors have been assumed (a total of 40,000 m²) and an additional 40,000 m² devoted to centralized photovoltaic plants, thereby using 0.2% of land identified as marginal by the US Geological Survey (1997). Figure 4.26 shows the time variations of produced electricity directly covering demands and of those used for hydrogen production. It is seen that fuel-cell uses of electricity are largest in the winter season but, in any case, such use is intermittent, with fairly few hours of high production. The heat loads (nearly constant except for January) are covered by the PVT panels, as indicated in Figure 4.27. A modest amount of associated heat from fuel-cell operation is required during the winter months during hours without solar production, but the waste heat created during summer has to be discarded.

The surplus electricity from centralized or decentralized photovoltaic plants is shown in Figure 4.28. This large export option could, of course, be reduced by installing fewer solar panels, but that would require more delicate use of energy stores, in addition to reducing an

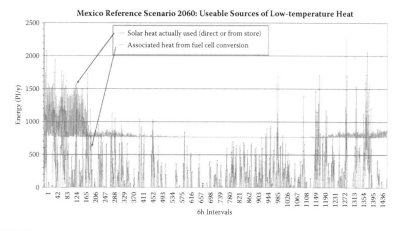

Figure 4.27   Coverage of Mexican demands for low-temperature heat by solar thermal installations (directly or through heat stores) or by associated heat from fuel-cell production of electricity, according to the 2060 reference scenario, as a function of time throughout a year and plotted
by 6-hour intervals.

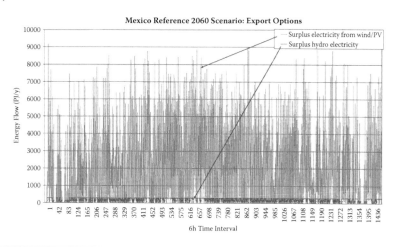

Figure 4.28   Potential electricity export options available to Mexico, according to the 2060 reference scenario, as a function of time throughout a year and plotted by 6-hour intervals.

obvious export opportunity (assuming that there are interested customers, e.g., in the United States).

4.1.6 Mexico with Doubled Electricity Demand

The variant scenario of a doubled electricity demand was considered
for the other North American countries. Although it may not be very
relevant in the context of Mexican development prospects, it could

Table 4.8   Summary of 2060 Reference Scenario with Doubled Electricity Demand for Mexico (PJ/y)

	LOW-T HEAT	ELECTRICITY	GASEOUS FUELS	LIQUID FUELS
Delivered energy demand	805	5088	1255	5477
Onshore wind-power production		846		
Offshore wind-power production		1169		
Hydropower production		285		
Photovoltaic power production		2563		
Biofuels from agricultural residues				1935
Biofuels from forestry residues				5101
Biofuels from aquaculture				720
Solar thermal energy produced	4777			
Electricity for dedicated uses		3735		
Electricity for hydrogen production		1105		
Liquid biofuels for transportation				5477
Liquid fuels used for power plants	(480)	600		1199
Hydrogen used in vehicles			444	
Hydrogen used in fuel cells	(42)	84	140	
Heat directly from solar panels	473			
Heat from local thermal stores	332			
Discarded or lost solar heat	3972			
Import requirement		669	671	0

Note: The heat energy values in parentheses are not used in the present scenario.

materialize if efficient use of electricity turns out to meet resistance in Mexico, despite its clear economic advantage (Figure 4.12). This would be analogous to the situation a few decades ago, when Mexico took over outdated US technology (e.g., cars) at low capital cost, albeit with high running costs, including that of the fuel for inefficient conversion. The high-demand scenario also has some theoretically interesting features that are explored in the following chapter to illustrate the technical functioning of regional trade arrangements.

Table 4.8 shows the summary characteristics of the Mexican high-demand scenario. Energy production is unchanged, but Mexico will change from an energy-exporting country to an energy-importing one. In this high-demand scenario, the transportation energy demand has been divided between liquid and gaseous fuels (15% of the latter) in order to strengthen the role of hydrogen as an energy-storage medium. This also leaves more biofuels for use outside the transportation sector,

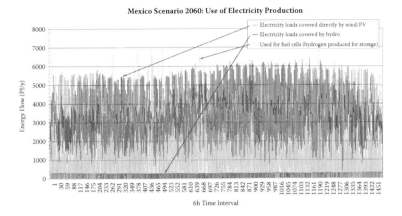

Figure 4.29   Mexican disposition of electricity produced by wind power, photovoltaic panels, or hydro energy for direct uses and for hydrogen production, according to the 2060 scenario with doubled electricity demand and onshore windswept area equal to 0.02% of land surface, as a function of time throughout a year and plotted by 6-hour intervals. The building-integrated solar collector area is 8 m^2 per capita and the central solar panel area 0.2% of marginal land.

notably in combined power and heat Carnot-cycle plants (CPH). The outcome of this is that more of the large electricity production away from hours of demand can be made useful, but there still is a deficit to be covered by import (now of both electricity and hydrogen), albeit much smaller than the electricity import need in the reference scenario. The time series of Mexican use of power produced is shown in Figure 4.29, and the time variations of the amount of hydrogen

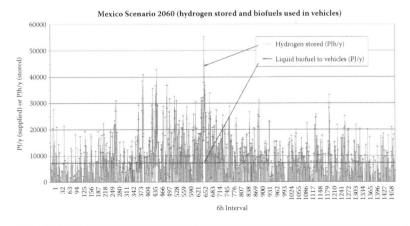

Figure 4.30   Mexican use of hydrogen storage and of liquid biofuels, according to the 2060 scenario with doubled electricity demand, as a function of time throughout a year and plotted by 6-hour intervals. The hydrogen store capacity is taken as 100,000 PJh/y, which constitutes less than 20 hours of average electricity usage.

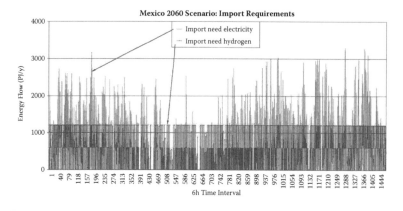

Figure 4.31 Required energy import (electricity and hydrogen) for Mexico, according to the 2060 scenario with doubled electricity demand, as a function of time throughout a year and plotted by 6-hour intervals.

stored is shown in Figure 4.30 along with the biofuel use. Finally, Figure 4.31 gives the time variations in import requirements for electricity and hydrogen. It is worth noting that the seasonal variations of the latter are small, suggesting that augmenting the hydrogen production (presumably by using more solar PV collectors) could reduce or eliminate the need for hydrogen import. Essential for such considerations are the storage options for hydrogen, because the cost of these will add to already-expensive photovoltaic panels and fuel cells. The following chapter continues this discussion.

References

Beijdorff, A., and P. Stuerzinger. 1980. Position paper on energy conservation. *11th World Energy Conference,* paper RT6. London: WEC.

Chow, T. 2010. A review of photovoltaic/thermal hybrid solar technology. *Applied Energy* 87:365–79.

ECMWF. 2013. European Centre for Medium-Range Weather Forecasts 40 Years Reanalysis (ERA40). Downloaded from http://apps.ecmwf.int/datasets/.

Melillo, J., and J. Helfrich, J. 1998. NPP database created under NASA and EPRI grants; based upon Melillo, J., A. McGuire D. Kicklighter, B. Moore-III, C. Vorosmarty, and A. Schloss. 1993. Global climate change and terrestrial net primary production. *Nature* 363: 234–240.

National Center for Atmospheric Research. 2006. Data Support Section. *QSCAT/NCEP Blended Ocean Winds ds744.4.* http://dssrda.ucar.edu/datasets/ds744.4/

Sørensen, B. 1982. Energy choices: Optimal path between efficiency and cost. In *Energy resources and environment*, ed. S, Yuan, 279–86. New York: Pergamon Press.

Sørensen, B. 1991. Energy conservation and efficiency measures in other countries. In *Greenhouse studies No. 8*. Canberra, Commonwealth of Australian Department of the Arts, Sport, the Environment, Tourism and Territories.

Sørensen, B. 2000. PV power and heat production: An added value. In *Proc. 16th European Solar Energy Conference, Glasgow*, 1848–51. London: James & James.

Sørensen, B. 2001. Modelling of hybrid PV-thermal systems. In *Proc. 17th European PV Solar Energy Conference, Munich*, 2531–34. Munich / Florence: WIP / ETA.

Sørensen, B. 2008a. A sustainable energy future: Construction of demand and renewable energy supply scenarios. *Int. J. Energy Research* 32:436–70.

Sørensen, B. 2008b. A renewable energy and hydrogen scenario for northern Europe. *Int. J. Energy Research* 32:471–500.

Sørensen, B. 2008c. A new method for estimating off-shore wind potentials. *Int. J. Green Energy* 5:139–47.

Sørensen, B. 2010. *Renewable energy: Physics, engineering, environmental impacts, economics and planning*. 4th ed. Burlington, MA: Elsevier.

Sørensen, B. 2011. *Life-cycle analysis of energy systems: From methodology to applications*. Cambridge, UK: RCS Publishing.

Sørensen, B. 2012a. *Fuel cells and hydrogen*. 2nd ed. Burlington, MA: Elsevier.

Sørensen, B. 2012b. *A history of energy*. New York: Earthscan/Routledge.

Sørensen, B. et al. 2001. Scenarier (in Danish). Final report for Danish Energy Agency project 1763/99-0001. Available as IMFUFA Text 390 (Roskilde University). http://rudar.ruc.dk.

US Geological Survey. 1997. *Global land cover characteristics data base*, v. 1.2. Earth Resources Observation System Data Center, University of Nebraska at Lincoln, Joint Research Center of the European Commission. http://edc2.usgs.gov/glcc/glcc.php.

Weizsäcker, E. von, C. Hargroves, M. Smith, C. Desha, and P. Stasinopoulus. 2009. *Factor 5: Transforming the global economy through 80% improvements in resource productivity*. London: Earthscan.

Weizsäcker, E. von, A. Lovins, and L. Lovins. 1997. *Factor 4: Doubling wealth—halving resource use*. Sydney: Allen & Unwin.

PART I
COOPERATION ACROSS AREAS AND REGIONS

The intermittency remedies discussed in Chapters 5–7 here in Part I comprise agreements for interchanging energy, whether in the form of trade agreements or other arrangements. Chapter 5 explores the opportunities offered by possessing or establishing power-grid connections by use of overhead or submerged power transmission lines, while Chapter 6 discusses the interchange of fluid or gaseous fuels by pipeline. Finally, Chapter 7 considers several other modes of trading energy, such as buying and selling fuels based on renewable or nonrenewable energy sources, using transfer other than the grid, notably by bulk transportation typically on ships or land vehicles, but conceivably also by air if economic factors support this option.

5

POWER-LINE INTERCHANGE

As discussed in connection with Figure 2.3, electric power supply systems comprise power production plants and lines transmitting power to areas of load (Figure 5.1), where transformers lower the voltage and feed power into a distribution network capable of reaching every single customer (Figure 5.2). Figure 2.4 indicates the options of connecting several such individual power supply systems by extending the transmission lines to enable transmission between the power systems. The cost of establishing such interconnections must be seen as viable in comparison with the benefits derived, such as increased system stability, insurance against the effects of inoperability of production units, and serving peak demands that would be expensive to satisfy with the production capacity available in a given, single system by use of less expensive units available in a neighboring power system, assuming that this is feasible, e.g., as a result of different time variations of loads in the two regions concerned. In cases where one or more of the interconnected systems employ variable energy sources such as wind or solar radiation, the grid exchange between systems can be an economically very attractive way of dealing with intermittency, despite the substantial cost of establishing new long-distance power lines (e.g., across the bodies of water) capable of transmitting in some instances the entire amount of electricity demanded at a given point in time in one region.

5.1 Present Use of Grid Interconnections and Near-Term Expectations

The use of grid interconnections for international exchange of power, or exchange between geographically separated utility networks within one nation, is already quite substantial, but such exchanges are used to varying degrees in different parts of the world. In Europe, cooperation by use of overhead or ocean cable transmission lines has several

Figure 5.1   Overhead transmission lines in Denmark (photo by author).

decades of history, and European Russia and the Asiatic parts of the country are connected by power lines that allow interchange between the several time zones. Less international interconnection is present in other parts of Asia (e.g., due to political concerns) and in North America. The absence of a common network operator may be one reason that collaboration between private utility companies appears to be less than optimal from the point of view of supply security. Recent debates on restructuring and privatization of the power utility

Figure 5.2   Transformer stations convert alternating current (AC) power from high-voltage transmission lines to the lower voltage of distribution grids (video still of facility in Tasmania). Other types of transformers are used to convert AC power to direct current (DC) for transmission through sea cables.

sector recognize that transmission and distribution are not amenable to competition (who would like ten competing lines to go into each building!) and thus should be controlled by a public or at least a non-profit body.

In South America, as in Africa, power interchange is mostly associated with common large hydro projects, serving to diminish the problems of fairly dividing access to water flows and reservoirs. Especially in Africa, the stage of development is such that many densely populated areas have not yet been electrified, and the several ambitious plans proposed for interconnections in all parts of Africa have remained unrealized plans, often due to the interruptions of collaboration caused by armed conflicts and genocides by ruthless dictators or by fundamentalist insurgence.

Figure 5.3 shows an outline of the current European electricity system (full details can be viewed at ENTSO-E [2013b]), and Figure 5.4 shows the average annual transfers among countries. The capacities of installed international connections (determining

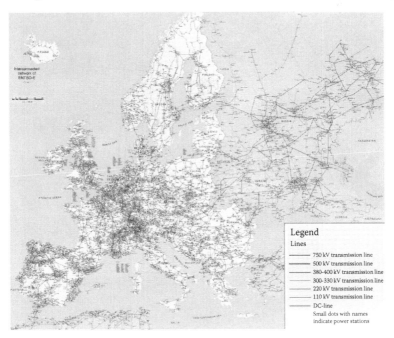

Figure 5.3 European electric transmission grid in 2012 (ENTSO-E [2013b]; used with generic permission). (This and several of the following network pictures are meant to illustrate the overall complexity of the grid structure; please disregard the abundant and often unreadable small text on several of the graphs.)

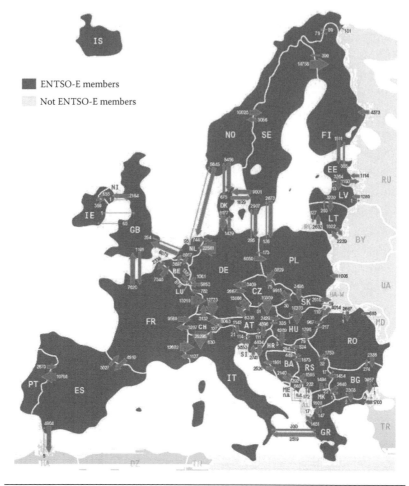

Figure 5.4   European power exchange, annual means in 2012 (ENTSO-E [2013c]; used by generic permission).

maximum transmission rates) are indicated in Figure 5.5. The avail-ability of power transmission lines shows considerable regional differ-ences. Some countries (Spain, Portugal, Italy, France, UK, Ireland, Iceland, and Poland) have import capabilities under 10% of their net generation capacities, and Germany, Greece, Bulgaria, and Romania are under 15%. The rest have more substantial transmission intercon-nections, notably the Nordic countries, where large elevated reservoir-based hydro shares in power production make the use of interchange desirable and profitable (e.g., for coping with "dry years," i.e., years with low precipitation, and as backup for any intermittency in the nonhydro power system) (Sørensen 1981; Meibom, Svendsen, and

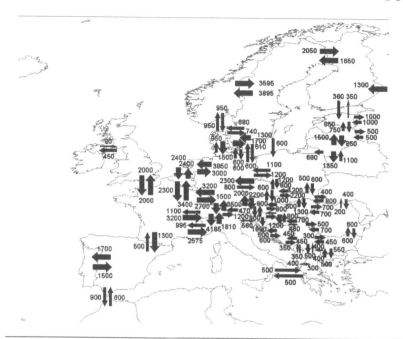

Figure 5.5 Indicative values for the international transmission line net capacities in continental Europe in 2010 (MW values based on data in ENTSO-E [2011]). The values are only indicative, because the maximum transmission through a given connection can depend on the status of the remaining (national) grid with respect to congestion and other factors. New interconnections keep being added to the system (such as lines between Norway and the Netherlands [700 MW] and between the Netherlands and England in 2012; cf. Figure 5.4).

Sørensen 1999; Schaber, Steinke, and Hamacher 2012; Doorman and Frøystad 2013; Jaehnert et al. 2014; Schroeder et al. 2014). Most European countries plan to expand transmission capacity during the coming decade (ENTSO-E 2013a; Kanevce, Mishkovski, and Kocarev 2013).

The arrangements for using the existing transmission facilities over land or sea are of three kinds: contractual fixed power deliveries at times convenient for the importing country, pool-based transfers based on daily auctions (administered by facilitators such as the Nordic company Nordpool), and finally agreed assistance in case of emergencies (such as the fallout of large generating units, e.g., nuclear plants). The capacity of individual transmission lines typically levels off at around 1 GW, and several parallel lines at that capacity level are preferred to the use of higher maximum power in one line. This ensures that the disturbance from a power line falling out is no greater than the failure of one of the larger power generation units. Handling

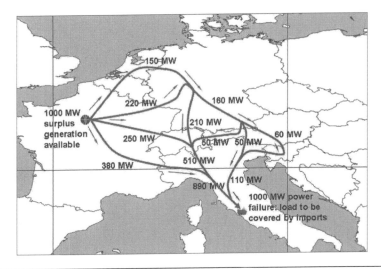

Figure 5.6 Transmission routing for an actual case of covering a 1-GW deficit in Italy by generation in France (based on data from OECD/IEA [2005] and Haubrich and Fritz [1999]). The roundabout transmission paths reflect the congestion on the European grid at the time in question (during the 1990s).

the falling out of a large generation unit often involves the use of many routes to the area missing power, including fairly roundabout itineraries (see Figure 5.6, based on an example discussed in OECD/IEA [2005]).

While most landlines carry alternating current at various voltage levels, the transmission system in northern Europe is characterized by extensive use of direct current lines for sea cables over considerable distances, with AC-DC/DC-AC converters at both ends (Nordel 2008). The possibility of using DC grid lines globally for transmission distances on the order of 10,000 km, with less than 2% power losses and acceptable economy, has been encouraged by Pickard (2014).

Formal collaboration between network operators across Europe has been in place for several decades and has been extended to include the non-European Union partners from the former Soviet Union. Expansion of transmission grids has progressed throughout the history of centrally produced electricity, on the basis of economic advantages or supply stability and resilience, but often with an extra motivation derived from being able to offer the convenience of electricity to as many citizens as possible, even at fairly high cost for peripheral regions, to be borne not by the new customers alone, but at least in some countries, shared with customers in the large load centers. From

Figure 5.7   Regional power transmission grid in former Soviet Union, 1964 (from Bar and Bater [1969]; reproduced with permission from John Wiley & Sons).

a business point of view, such a policy may be warranted, because a grid extension that is not instantly cost effective may become so as the new areas being supplied develop their population and energy-consuming activities. The founder of the Soviet Union, Lenin, often emphasized the importance of electricity and may have encouraged funding of power transmission projects, but still, the vision of avoiding the need to size power plants to supply peak loads by instead using transmission from one time zone to the next (assuming peaks to fall at the same local time) did not materialize in the Soviet Union until the late 1950s (Figure 5.7) (Thiel 1951; Barr and Bater 1969; Cigré 2011). A proposed European–Asian network cooperation would cover 13 time zones (by combining the European network of Figure 5.5 with the northern part of the Asia grid shown in Figure 5.8, as suggested by Bollinger-Kanne [2008]). The dynamics of interconnections over the several Siberian time zones have been modeled by Bondereva et al. (2004).

In North America, within countries and between the United States and Canada, the ratio of transmission capacity between grid operating entities and their maximum loads tends to be a little smaller than in Europe, and between the United States and Mexico it is much smaller. The main international connections are between the upper west coast and east coast US states and Canada, with its extended

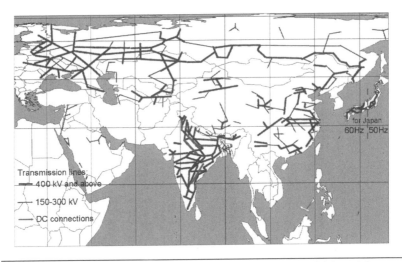

Figure 5.8 Principal power transmission grids in selected countries of Asia and European Russia, in the early twenty-first century (with use of data from State Grid Corporation of China [2013]; Federation of Electric Power Companies of Japan [2013]; Bollinger-Kanne [2008]; Meissen and Mohammadi [2010]; GENI [2013]; International Energy Agency [2002]).

reservoir-based hydro utilization (Canadian Electricity Association 2002, 2010). Figure 5.9 shows the US grid, and Figure 5.10 shows its capacities for international transmission through interchange with Canada and Mexico. Several extended blackouts in recent years have spurred activities in the United States to improve the strategies for

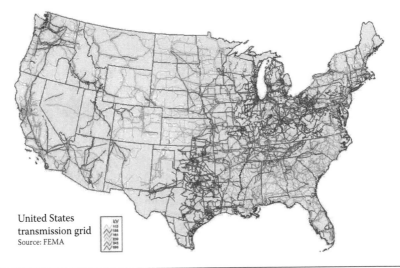

Figure 5.9 US transmission grid (115–500 kV, 2008). (Wikimedia Commons, public domain picture created by J. Messerly.)

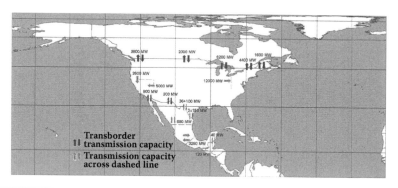

Figure 5.10 Power line capacities for transmission between countries in North America at about 2010 (MW, black arrows), and some indications of maximum transfer capacities (MW, gray arrows) between regions in one country, across the dashed lines. The capacity estimates are approximate, because the current that can be transmitted through a line of given voltage and structure depends on the state of the grid and on meteorological conditions (Silverstein [2011]; Mexican Department of Energy [2010]; EIA [2011]).

dealing with congestion and emergencies (from natural disasters to power plant fallout; see Hauer et al. [2002]). Figure 5.11 presents the relationship between outage periods and magnitudes, showing a trend toward larger load losses in each instance, based upon comparing year 2000–2004 data with 1990–1994 data (OECD/IEA 2005). Outage durations range from minutes to some 40 days, and the largest amount of power loss is close to 50 GW, out of a total US generation capacity of about 1000 GW (EIA 2010).

In Mexico, the grid (Figure 5.12) allows interchange between several of the country's load regions (Mexican Federal Electricity Commission 2011), and has a very modest international transmission capacity (Figure 5.10) for reaching the chain of interconnected power grids in Central America (Figure 5.13). Here, there is a partially completed backbone of transmission lines all the way along the West coast of Central America, but very few lines along the eastern coast and lowland areas (Reinstein et al. 2011; Central American Electrification Council 2012). Obvious interconnection possibilities between Florida or Yucatan and Cuba and between Cuba and Jamaica have not been considered, perhaps for political reasons. (However, in other cases, such as the earlier transmission collaboration between the Soviet Union and central or northern Europe, efforts to optimize the electricity supply and political differences have been decoupled, at least to a certain extent.)

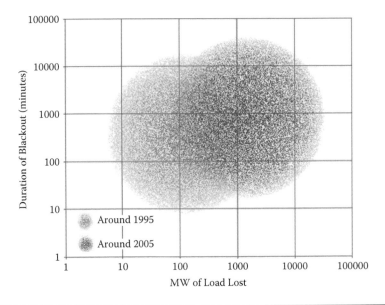

Figure 5.11 Trend in development of size-duration characteristic of power outages from 1995 to 2005, based on a stochastic model with parameters adjusted to the US case study considered by OECD/IEA (2005). By 2005, the number of disturbances has increased, as has their size and to some extent their duration. Whether this trend has been reversed by measures taken in recent years to reinforce transmission facilities remains to be seen.

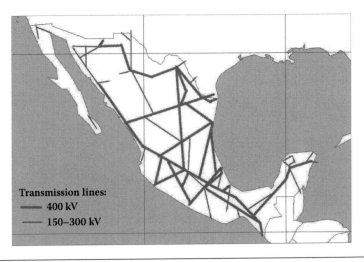

Figure 5.12 Overall Mexican power grid in 2011, with extensions under construction to 2014 (with use of data from Mexican Federal Electricity Commission [2011]).

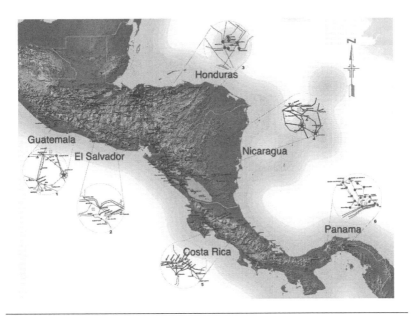

Figure 5.13 Central American transmission grid in 2004, with expansions envisaged (but not all realized) for 2006 (Costa Rican Institute of Electricity [2004]; used with permission).

Several regions far from the urban areas of South America have weak transmission networks (Madrigal and Stoft 2012). This is the case for the southwestern part of Venezuela, as it is for the northwestern part of Brazil (Figure 5.14). The connection between the two countries at the moment consists of a detached line between a power station in Venezuela and a load area in Brazil, far from the main grid networks of either country. Brazil has a large amount of hydropower and a sizeable transmission network covering the large urban regions. International connections include smaller lines to Uruguay as well as two substantial 500-kV transmission lines to hydro facilities in Argentina and Paraguay. Longer overhead transmission lines in the southeastern part of the country have suffered from failures caused by atmospheric discharges and similar environmental factors (Duro et al. 2012). The grid in Argentina (Figure 5.14) contains an international link to northern Chile, the mentioned large one to Brazil, and smaller connections to Uruguay and Paraguay. Only recently has the transmission line reaching the southern parts of the country been installed. As regards Chile (also Figure 5.14), a substantial transmission facility exists in the central parts of the country, as well as in the north, but the regions are not connected, and the far south is served by several

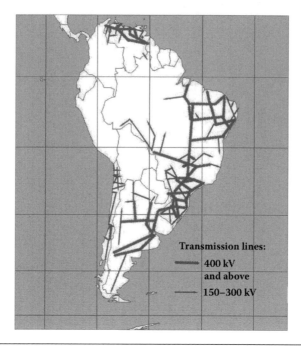

Figure 5.14 General electric power grid in selected South American countries, at about 2010 (with use of data from Corpoelec Venezuela [2013]; Brazilian Operator of the National Electricity System [2012]; Argentine Federal Electricity Council [2011]; Chilean National Energy Commission [2005]).

small hydro plants without interconnection and by other small independent producers.

New Zealand transmission grids reach most populated areas, and the North and South Islands are connected by a 350-kV direct current cable (Figure 5.15). Most of Australia is without perceived need for power grids, which are found only near the populated regions in the southeastern coastal regions and around Perth on the west coast (Figure 5.16). Prospects for populating the inland areas are bleak due to the island's general shortage of water. Possible solutions could be solar desalination of ocean water plus long-distance pipeline transport (as practiced in areas of a similar climate in Spain), or more elaborate manipulations of the water cycle by seeding clouds and building structures to reduce evaporation (such as cities totally covered by solar-panel domes reflecting the energy not harvested or used for lighting). Electric interconnection of the Darwin area in the north with the heavily populated Southeast Asian countries could prove beneficial in allowing export based on the substantial renewable energy power

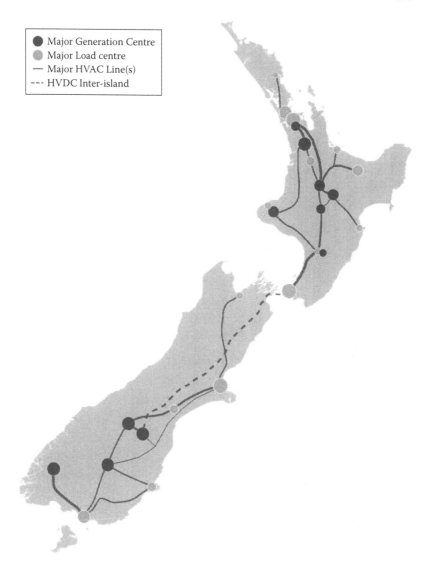

Figure 5.15 Outline of New Zealand power grid in 2009, with indication of power stations and load centers. The dashed line denotes DC cable (LC Mortensen, Wikimedia public domain picture).

options available in (all of) Australia. However, at present, there is no connection from Timor (a possible undersea cable endpoint some 300 km from Australia) to or along the string of more-populated Indonesian islands (except for the connection between Java and Bali), so several initiatives have to be taken if Indonesia's current problems of frequent blackouts are to be solved by international collaboration (Reuters 2008).

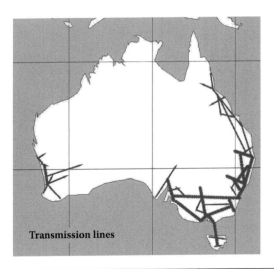

Figure 5.16   Power transmission grids in Australia in 2009 (with use of data from Grid Australia [2013]).

China has been very active in adding power generation capacity and transmission lines during its growth period after the downturn caused by Mao's Cultural Revolution. Still, there are extensive areas with little transmission infrastructure (Figure 5.8), and the rapid transformation of the country with steeply increasing power demands poses big challenges for further development of the electricity system in the near future. The Japanese power grid (at the far right of Figure 5.8) reflects the division of the country between several private electricity suppliers with only modest collaboration, highlighted by the continued reliance on two different frequency standards in the eastern and western parts of Japan. There are only weak transmission links to the northern provinces and no international interconnections. These features and particularly the modest capacity for frequency conversion (that would have allowed west-to-east transmission to the affected areas) played a substantial role in the supply cuts and electricity rationing becoming necessary in the wake of the 2011 Fukushima nuclear accident. This association and its implications for selecting the sources of electricity generation and the future network structure seem to be ignored in some of the many scenario discussions regarding the future development of Japanese energy capacity (Pereira, Parady, and Dominguez 2014).

Figure 5.8 also shows the power grid in India, which is interesting due to its use of direct current for transmission over land, with

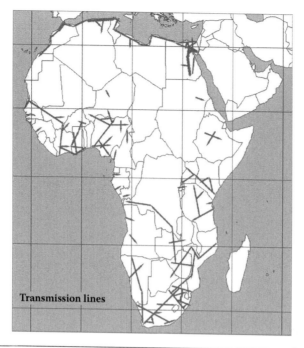

Figure 5.17   Overall electric power transmission grid in Africa, early twenty-first century. A large number of unrealized planned lines announced over the past 20 years have been omitted (based on data from Egyptian Ministry of Electricity and Energy [2013]; GENI [2013]; Kinyanjui, Gitau, and Mang'oli [2011]).

several low-loss thyristor facilities for AC-DC or DC-AC conversion. In Europe, such facilities have, until recently, primarily been used in connection with undersea cables (Messerly 2013). The east-central Asian region stretching from the Caspian Sea to China (Figure 5.8) has a well-developed transmission grid except for low-consumption areas, and there are several interconnections between the countries and to Russia, originating in the structure of the former Soviet Union.

The same cannot be said for Africa, despite some areas having developed a power infrastructure and hoping to improve it, such as South Africa and several coastal city areas, plus the lower Nile area, as shown in Figure 5.17 (Eberhard et al. 2011; Kessides, Bogotic, and Maurer 2007; Egyptian Ministry of Electricity and Energy 2011). A large number of plans and proposals for enlarging the transmission grids have emerged over the past 20 years, but hardly any have been implemented in a continent ridden by conflicts, dictatorial rulers and, in recent years, frequent insurgence by popular movements

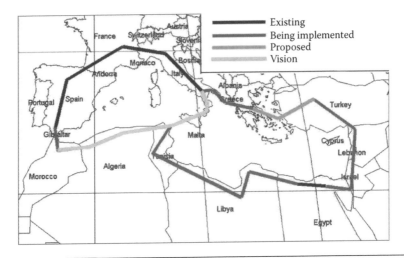

Figure 5.18 Proposed Mediterranean Ring grid aimed at strengthening the North African grid and eventually allow Europe to import power from Sahara Desert solar plants (Sørensen [2010]; used with permission).

demanding either more democracy or more fundamentalist assaults on basic human rights. A number of international network associations have been formed, such as the East African one responsible for ambitious visions of interconnectivity (East African Community 2008, 2009; Kinyanjui, Gitau, and Mang'oli 2011), but again proposals have remained proposals.

Figure 5.18 shows the ring of transmission lines around the Mediterranean Sea, which have repeatedly been discussed, notably by the former Egyptian minister for energy (Abaza 1996) and by several international organizations (IEA PVPS 2013; Desertec 2013) who see the enormous potential for centralized photovoltaic or concentrating solar thermal plants in the little-populated Sahara region, with obvious export options to both African and European countries (Komoto et al. 2013). Only the connection across the Gibraltar Strait has actually been installed, and it is currently used primarily for export of power from Spain to Morocco (cf. Figure 5.4), in line with the limited use envisaged by Brand (2013).

There is still what may be interpreted as a limited understanding of the detailed relationship between the intermittency of certain renewable energy sources and their abilities to match demands. Often, the influence of system size, including the number of dispersed generating units as well as the transmission network layout and strength, of geography (variability of wind and solar influx with location), and of system management (such as a cooperative environment versus a competitive one) are not taken fully into account. This is particularly true for actors from the economic sector (for an example, see Madrigal and Porter [2013]), who tend to focus on direct costs, whereas actors far from the economic sector tend to either totally disregard cost or at best insist on basing decisions on complete life-cycle direct and indirect costs. The actual cost of the recent Norwegian underwater cables to the Netherlands (700 MW) and Germany (1400 MW), both of lengths over 500 km, is estimated to be about 0.5 and 1.2–2.7 million euro per km, respectively (Chatzivasileiadis, Ernst, and Andersson 2013; Energinet 2013). A positive example of recognizing the relation between access to transmission and integration of renewable energy may be found in Tawney, Bell, and Ziegler (2011). Efforts to tune down cost arguments are common among advocates for global transmission grids (essentially just drawing any number of straight lines between continents) independent of need (GENI 2013; Chatzivasileiadis, Ernst, and Andersson 2013), but of course such criticism can always be countered by suggestions of radical new technological breakthroughs that fundamentally alter present cost estimates. One such technology could be superconducting power transmission, so far used only in scientific laboratories and prototype demonstration projects (Masuda, Yumura, and Watanabe 2008; Sim et al. 2008). Bulk power-sector uses appear today to be associated with extremely high but very uncertain cost perspectives.

5.2 Impacts Associated with Limited Transfer Capacity

The geographical survey presented in Section 5.1 indicates several regions where the transmission capacity is limited. This is partially decoupled from the issue of electrification. For example, rural electrification may be achieved by use of decentralized and unconnected generators. Reliability can be ensured by suitable arrangements, such as providing battery storage for use during generator breakdown and repair, or backup arrangements such as adding a fuel-based gas turbine to a photovoltaic system. However, these solutions are economically warranted only if the cost of interconnected grid systems is higher than that of the dispersed systems. In most cases, interconnection by transmission networks is by far the most economic solution whenever the load density is substantial (which would translate into a condition on population density and level of welfare). However, there is an evident dependence on the distance to the other load centers to which interconnection is made as well as on the peak level of power transfer deemed necessary.

For some types of electricity sources, there is a geographical dependency not related to the density of demand. This is the case for large hydro installations, which are necessarily restricted to the locations where elevated water is present, often in mountainous regions far from load centers. This is not in all respects a disadvantage, because it diminishes the impact on existing human settlements connected with reservoir establishment (but usually not the impacts on the natural environment). However, there are examples of potential hydro projects that have been abandoned or limited in size due to their distance from load centers. In South America and Africa, new hydro projects—each allowing generation of several GW of power—are possible, but only some have been realized, because in the other cases, no load areas of corresponding size were present at distances deemed reasonable for the establishment of transmission lines.

As elaborated on toward the end of the previous section, the relationship between transmission options and the introduction of large-scale but dispersed solar and wind power is quite complex. A key factor is of course the kind of other generating units that are connected to the transmission grid, a consideration that hinges on whether they are hydro or fuel/biofuels plants, the output of which can usually be adjusted over quite large intervals, or other variable sources without immediate output control, such as more wind and solar energy. In the renewable-flow cases, the important thing is whether these power generators are placed sufficiently dispersed to experience different climatic environments. For wind power, the short-term variations are smoothed out when several plants are attached to the same network, even at small mutual distances (under 100 km), whereas the variations caused by passage of weather fronts usually have an extent of several hundred kilometers, so that smoothing the combined output would require interconnection over distances of 500 km or more (Sørensen 2010). The same is true for photovoltaic panels, except that here the climatic variations are fairly smooth (basically latitude dependence), and most variations are due to cloud cover (and of course day and night). As mentioned previously, systems covering several time zones offer additional advantages, although they cannot make up for the basic variations (due to length of day and changes in incident angle). Current international connections, even in the leading Nordic European countries, are still under 50% of the power demand peak in the country with most wind energy (Denmark), and in many other regions it is under 10%.

Finally the role of transmission in stabilizing power supply should be mentioned. This is important for all energy sources, but more so for large fuel-based, hydro, or nuclear plants, because dispersed wind or solar installations are unlikely to fail simultaneously. A sudden fallout of a large generation unit not only calls upon transmission to furnish the lost power, but also may create uncontrolled oscillations in the network around the failed power plant. Some resilience against this kind of instability may be achieved by suitably designed parallelism in the network structure, and this is the subject of intense investigations of both theoretical and practical nature, aimed to prevent failures from cascading to other sites on the grid (NERC 2013; Koç et al. 2013; Carreras et al. 2002; Hauer et al. 2002; OECD/IEA 2005; Kinney et al. 2005). An example of complex grid relief supply was discussed in connection with Figure 5.6.

5.3 Advanced Use of Power Exchange for Handling Intermittency

In addition to helping cope with power disruptions, the role of transmission in bringing power to load centers from sites where construction of power plants is particularly attractive has been mentioned previously, notably as regards large hydropower stations. However, for variable resources such as wind and solar energy, the use of transmission lines can be an even more fundamental part of the concept of demand–supply matching, inviting the establishment of quite long international grid connections. In particular, the availability of hydro energy storage facilities with seasonal reservoirs (or, for that matter, any other large-scale energy storage) can allow nearly 100% coverage of demands with the variable renewable sources, as first suggested by Sørensen (1981) in a study combining Danish wind power with Norwegian hydro storage. Figure 5.19 shows the monthly transfers through expanded

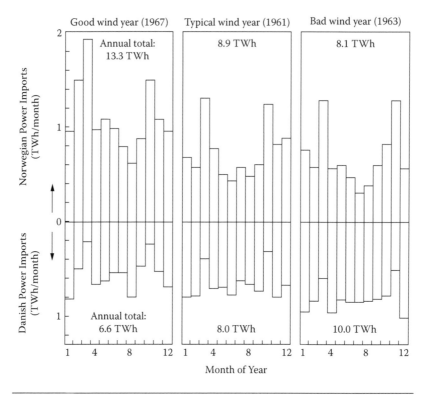

Figure 5.19 Monthly average results of an hourly simulation of transmission requirements in one or the other direction for a Danish–Norwegian wind–hydro system, using data from three different Danish wind years. The interconnection capacity of 5 GW is sufficient to transmit the deficit or surplus power at any time (Sørensen [1981]; used with permission).

transmission connections for a good, an average, and a poor wind year, thereby indicating the effect of interannual variations in wind energy resources. The study assumed the then-existing (1980) Norwegian hydro system and a hypothetical 100% wind-power-based system in Denmark. The Danish deficit periods covered by Norwegian hydro-power during the average wind year (8.9 TWh/y) constituted 41.5% of the Danish load, but the Norwegian hydro production was not found to be affected, because the power "borrowed" by Denmark is paid back from surplus Danish wind power within 2–3 weeks, causing less than 2% change in the reservoir filling in Norway (during both good and bad Norwegian years in terms of precipitation and reservoir filling), and furthermore, the borrowed power can be produced by the already-existing surplus capacity of the then-operating Norwegian hydro turbines. This scenario requires a total transmission capacity between

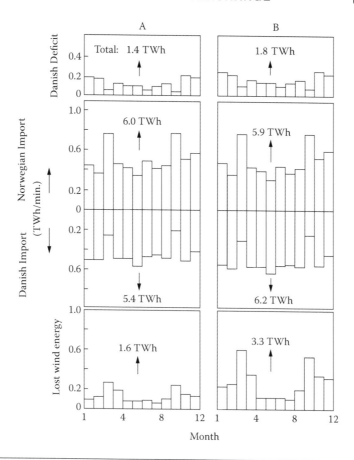

Figure 5.20 Monthly average results of an hourly simulation of the same system as in Figure 5.19, except that the interconnection capacity equals the actual one (1.5 GW) in 1980. Ensuing Danish power deficits are shown at top, and surpluses over the assumed transmission capacity at the bottom (Sørensen [1981]; used with permission).

Denmark and Norway (directly or through Sweden) of 5 GW, or 3.3 times the actual capacity in 1980.

In the alternative simulation behind Figure 5.20, the transmission capacity has been restricted to the one actually in place by 1980, showing that now some wind energy cannot be made useful, and there will be occasional deficits that have to be covered by other means, such as natural-gas-fired power plants. In this calculation, Denmark will have a deficit of 1.4 TWh/y (6.5%) to be produced by sources other than wind, and a 1.6-TWh/y wind production that cannot be transmitted to Norway and will be lost if no other option is considered (such as export to Germany).

Two studies mentioned in Section 5.1 (Meibom, Svendsen, and Sørensen 1999; Jaehnert et al. 2014) address the same problem from an economic point of view. On the one hand, trade of electricity must follow the rules currently set for the auction arrangements ("power pools"), and on the other hand, investments in new transmission capacity must reflect an expectation of long-term economic viability. The current operation of the Nordpool power-exchange facilitator is tailor-made to power generators based on freely adjustable output and nonvariable energy sources. It requires that bidding take place 36 hours before the actual exchange, reflecting a perceived computer illiteracy among the hydropower operators that may not prevail. In other pool systems such as the Australian one, bidding and actual trade is virtually instantaneous. This would of course be highly appropriate when the reservoir-based hydro is used to smooth variable renewable inputs, and the extended role of Norwegian hydro as a complement not only to Danish wind, but to European photovoltaic and wind power (e.g., in Germany and the Netherlands) may change the attitude toward trading rules (Jaehnert et al. 2014). The earlier very small price paid for electricity in Norway is already approaching European average prices, as it will have to, once the producer has the choice of selling to local customers or exporting this valuable merchandise, albeit having to incorporate the cost of suitable transmission capacity.

The simulations made in Meibom, Svendsen, and Sørensen (1999) resemble those of the 1981 study, except that now the region is divided into eight sectors that are analyzed in terms of demand, production, and transmission (Figure 5.21), and the emphasis is on exploring the ability to forecast wind power production by the 36 hours required by current Nordpool rules, using advanced extrapolation or general circulation prediction software. The conclusion is that the mean error of wind power predictions for pool bidding is on the order of 15%, which may or may not be a problem, depending on the penalty that the Nordic market claims for making errors. Hopefully, the future will see changed bidding rules that make this concern irrelevant.

A more general northern European model with renewable energy and substantial trade and transmission has been constructed by Sørensen (2007), comprising not just electricity, but all energy forms demanded. In addition to the large, reservoir-based hydro in Sweden and Norway, there are vast forestry residue resources that may be

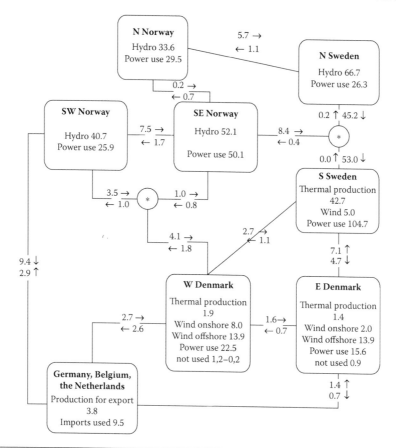

Figure 5.21 Model structure and summary of simulation results for a Danish–Norwegian 2030 model with transmission to other European countries and only renewable power generation in Denmark (from wind and biomass, the latter called *thermal production* in the picture). The size of production, load, and transmission is given in TWh/y (Meibom Svendsen, and Sørensen [1999]; © Inderscience, used with permission).

recovered in a sustainable way, particularly in Sweden and Finland. This defines three Nordic countries with options for substantial exports, plus Denmark, where renewable resources roughly balance demands, and Germany, where the renewable resources (chiefly sun, wind, and biomass) appear insufficient for covering future demands, particularly in the transportation sector. The scenario work identifies stable solutions to matching demand and renewable supply, either by north-to-south export of biofuels or by export of electricity, which after arrival (in Germany) can be used to generate hydrogen for transportation uses. Time-simulation results for the hydrogen scenario are shown in Figure 5.22. The required transmission capacity to Germany

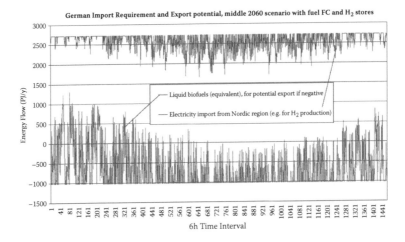

German Import Requirement and Export potential, middle 2060 scenario with fuel FC and H$_2$ stores

Figure 5.22   Possible 6-hourly 2060 coverage of German deficit in indigenous renewable energy power and biofuels production, by import of surpluses in the Nordic countries, according to one scenario discussed in Sørensen (2007; used with permission). Note that for biofuels, Germany has a surplus for export during many hours over the year. The reason is that hydrogen is assumed to be the preferred energy source in the transportation sector, and that the hydrogen production requires substantial use of imported electricity.

is as high as 86 GW, but could be reduced if an economic comparison indicates that it would be less expensive to expand the capacity of the hydrogen reservoirs (which in the scenario are small) and suffer the ensuing losses in the electricity-to-hydrogen-to-electricity cycle. The alternative of liquid biofuel import would comprise an average transport of 102 GW of transportation fuels.

5.3.1 A High-Transmission Scenario for North America

The case studies presented in Chapter 4 offer the possibility of further exploring the options for power transmission and other forms of import/export, considering other regions of the world than Europe. Here, a North American scenario will be developed, aimed at dealing with the problems of the high-energy-demand scenarios without interconnected regions as presented in Sections 4.1.2 to 4.1.6. The largest potential electricity production surpluses for export are found in the northern areas of Alaska, Canada, and Greenland. Alaska and Canada also have the possibility of producing surplus biomass for export of biofuels. The United States is roughly self-sufficient in renewable electricity supply (with a small surplus for export, despite

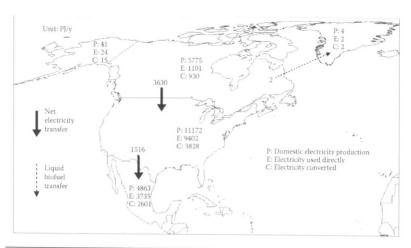

Unit: PJ/y

P: 41
E: 24
C: 15

P: 4
E: 2
C: 2

P: 5775
E: 1101
C: 930

3630

Net
electricity
transfer

P: 11172
E: 9402
C: 3828

1516

P: Domestic electricity production
E: Electricity used directly
C: Electricity converted

Liquid
biofuel
transfer

P: 4863
E: 3735
C: 2601

Figure 5.23 North American electricity transmission scenario for an all-renewable energy supply. Numbers given are annual average transfers, and the time variations and grid requirements are discussed in connection with Figures 5.24 and 5.25. Transmission losses are taken as 5% for each connection. The 3630-PJ/y transmission capacity required between Canada and the United States translates to 115 GW.

the assumed doubling of indigenous demand), but lacks large amounts of hydrogen fuel for its transportation sector. Mexico has a deficit of both electricity and hydrogen fuel, and Greenland would have to import its (tiny) biofuels demand or drive only hydrogen vehicles.

One scenario solving the mismatch problems of a future North America based on renewable energy—including intermittent sources—would employ electric power transmission from north to south, sufficient to cover deficits and allow the receiving countries to generate their own hydrogen fuel by electrolysis from the imported electricity plus the indigenous amounts already included in the scenarios of Tables 4.1–4.8. Figure 5.23 shows the way in which this can be accomplished for the scenario where electricity demand is doubled relative to the reference minimum: In this scenario, Mexico has a 13% deficit of electricity and a hydrogen deficit greater than 50% (Table 4.8). The United States has a small surplus of electricity but not enough to cover the Mexican deficit. It is therefore necessary to bring some of Canada's huge potential electricity surplus south to Mexico, both for direct use and for producing hydrogen by electrolysis. However, the United States also has a two-thirds deficit of hydrogen, so the amount of Canada-to-US power transmission in Figure 5.23 is taken to cover all hydrogen deficits in the United States and Mexico, plus the small Mexican deficit in direct electricity not covered from the United States, plus the transmission and conversion losses along the way. However, the actual transmission of electricity would, of course, not be directly from Canada to Mexico, but only to the closer demand locations in the United States, from where some locally produced power can now be transmitted further South instead of covering local demands, and so on. It is considered to be less expensive to transfer electric power and do the hydrogen conversion at the receiving end, rather than to doing it in the country of surplus origin and then transferring hydrogen by pipeline.

To evaluate whether this is the best strategy, the time variations in power available for transfer and in the demand of the importing country must be checked. Figure 5.24 shows the time dependence of the Canada-to-US transmission, and Figure 5.25 shows that of the US-to-Mexico transmission. It is seen that the transfers vary substantially and that they are sometimes reversed relative to the overall north–south direction. This is due to occasional lack of surplus in the Canadian system, typically at times of low wind, while high wind is

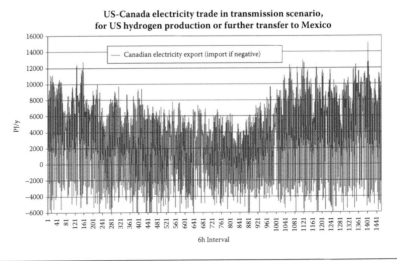

Figure 5.24 Time sequence of positive or negative power transmission from Canada to the United States in the North American transmission scenario. Although the variations chiefly illustrate variations in Canadian wind energy, the presence of reservoir-based hydro in Canada would make it possible to smooth the curves and avoid power imports into Canada. Also on the receiving end, much of the electricity imported by the United States is used to generate hydrogen, which is not a time-urgent application. The transmission capacity between Canada and the United States could thus be reduced, say to about 250 GW, without problems (full timewise smoothing would make 115 GW suffice if imports were constant).

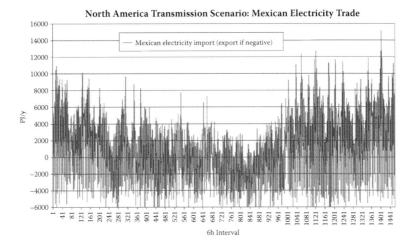

Figure 5.25 Time sequence of positive or negative power transmission into Mexico from the United States, according to the North American transmission scenario. As in Figure 5.24, a considerable part of the power imports is used to generate hydrogen, which would allow reducing or avoiding the negative imports (exports) present due to the assumption of constant Mexican hydrogen production over the year. The capacity of the transmission lines could thereby be somewhat reduced from the peak import value (some 440 GW), say to about 250 GW (full smoothing would make 50 GW suffice, but that would require Canadian exports to be kept constant by use of the Canadian hydro facilities).

the main cause of the occasional large surpluses. The variations could therefore be diminished if extra Canadian hydro was used for domestic supply during low-wind periods and correspondingly less hydro in periods of wind energy surplus. In this way, the overall functioning of the hydro reservoirs is not affected (similar to the situation in the northern European model described previously). Further control of variations can be achieved at the receiving end by varying hydrogen production in the United States and in Mexico according to the availability of power import, rather than the assumption used in the model of taking it to be constant over the year. Such measures would allow the US-Mexico power transmission variations in Figure 5.25 to be similarly reduced. Using Canadian hydro to smooth the import requirement would further allow the transmission capacity to be lowered, both from Canada to the United States and from the United States to Mexico. In addition, allowing variable hydrogen production would avoid or reduce the reverse (south to north) transmission and thereby diminish the peak import values, but not eliminate peaks altogether, precisely because the electrolysis units now run at the variable output determined by the variations in available power for import. The Canadian hydro production assumed in the model (1170 PJ/y, which is similar to the current one, see Table 4.3) is not sufficient to totally flatten the power export over the year.

Whether this power-transmission-based scenario for North America is attractive or not depends on the cost of the transmission lines required, as compared to other ways of matching demand and supply. The cost of a 700-MW sea cable quoted in Section 5.1 is consistent with other estimates of about 1 million 2014-euros per kilometer (Alegría et al. 2009), but that of transmission by overhead lines over land is 5 to 10 times lower per kilometer, again depending on additional factors such as land prices, transmission distance, and parallelism in the grid layout (Niederprüm and Pickhardt 2002). Silverstein (2011) quotes a cost of US$0.55 million/km for a 750-MW landline in the United States and $1.4 million/km for a 3000-MW line. The cost of hydrogen pipelines and of biofuels transport over land or by sea should be compared with transfer using electricity lines.

References

Abaza, A. 1996. Pan-African–Asian–Europe regional electric interconnection. In *Proc. 3rd Afro-Asian Int. Conf. on power development strategies for 21st Century*, Kathmandu, Nepal. Vol. 1, 3–11. New Delhi: Oxford & IBH Publishing.

Alegría, I., J. Martin, I. Kortobarria, J. Andreu, and P. Ereño. 2009. Transmission alternatives for off-shore electrical power. *Renewable and Sustainable Energy Reviews* 13:1027–38.

Argentine Federal Electric Energy Council. 2011. Map of transmission lines available from the Consejo Federal de la Energia Eléctrica, Buenos Aires. http://www.cfee.gov.ar.

Barr, B., and J. Bater. 1969. The electricity industry of central Siberia. *Economic Geography* 45 (4): 349–69.

Bollinger-Kanne, J. 2008. Von Wladiwostok bis Lissabon. *Energiespektrum*, no. 5:30–31.

Bondereva, N., D. Kolotovkin, R. Cherkaoui, A. Germond, A. Grobovoi, and M. Stubbe. 2004. Comparison of the results of full-scale experiment and long term dynamics simulation in the Siberian Interconnected Power System. In *IREP Symposium Proc.: Bulk power system dynamics and control*, Cortina-d'Ampezzo, Italy. Vol. 6, 492–98.

Brand, B. 2013. Transmission topologies for the integration of renewable power into the electricity system of North Africa. *Energy Policy* 60:155–66.

Brazilian Operator of the National Electricity System. 2012. Power transmission maps. http://www.abradee.org.br or http://www.ons.org.br.

Canadian Electricity Association. 2002. *Developing a North American energy perspective.* Canadian Electricity and the Economy Series. Ottawa: CEA.

Canadian Electricity Association. 2010. *The integrated electricity system.* Ottawa: CEA.

Carreras, B., V. Lynch, I. Dobson, and D. Newman. 2002. Critical points and transitions in an electric power transmission model for cascading failure blackouts. *Chaos* 12 (4): 985–94.

Central American Electrification Council. 2012. *Plan Indicativo Regional de Expanción de la Generatión. Perioda 2012–2027.* Working group paper from Consejo de Electrificatión de América Central, Panamá. (Additional material is in the anniversary edition of the plan for 2011–2025 [2010].)

Chatzivasileiadis, S., D. Ernst, and G. Andersson. 2013. The global grid. *Renewable Energy* 57:372–83.

Chilean National Energy Commission. 2005. *La Regulatión del Segmento Transmisión en Chile.* Working paper from Comisión National de Enérgia, Santiago. http://www.cne.cl.

Cigré. 2011. *The History of Cigré.* Paris: International Council on Large Electric Systems.

Corpoelec. 2013. Power transmission map of Venezuela available from the government network operator. http://www.corpoelec.gob.ve.

Costa Rican Institute of Electricity. 2004. *Power in Latin America.* San José: Instituto Costarricense de Electricidad ICE.

Desertec Foundation. 2013. Desert PV solutions. http://www.desertec.org.

Doorman, G., and D. Frøystad. 2013. The economic impacts of a submarine HVDC interconnection between Norway and Great Britain. *Energy Policy* 60:334–44.

Duro, M., P. Kaufmann, F. Bertoni, E. Rodrigues, and J. Filho. 2012. Long-term power transmission failures in southeastern Brazil and the geophysical environment. *Survey Geophysics* 33:973–89.

East African Community. 2008. *EAC strategy to scale-up access to modern energy services.* Country reports for Kenya (A. Ngigi), Tanzania (F. Magessa), Rwanda (J-C. Nkurikiyinka), Uganda (M. Okure). Arusha, Tanzania: EAC.

East African Community. 2009. *Regional strategy on scaling-up access to modern energy services in the EAC.* Arusha, Tanzania: EAC.

Eberhard, A., O. Rosnes, M. Shkaratan, and H. Vennemo. 2011. *Africa's power infrastructure.* Washington, DC: World Bank.

Egyptian Ministry of Electricity and Energy. 2011. *Egyptian Electricity Holding Company annual report 2010/11.* Cairo: Arab Republic of Egypt.

Egyptian Ministry of Electricity and Energy. 2013. Transmission map. Cairo: Arab Republic of Egypt. http://www.moee.gov.eg.

EIA. 2010. Generation statistics. US Energy Information Administration. http://www.eia.gov/electricity.

EIA. 2011. *Today in Energy.* 12 December. US Energy Information Administration. https://www.eia.gov.

Energinet. 2013. Website of the publicly owned Danish power-network operator. http://www.energinet.dk.

ENTSO-E. 2011. *NTC-values winter 2010–2011*. Spreadsheet. https://www.entsoe.eu under "NTC-matrix" created 24.02.2011. (ENTSO-E is an association of European network operators working with the European Union. Some newer country-related data may be found at www.entsoe.net.)

ENTSO-E. 2013a. *Ten-year network development plan 2012*. https://www.entsoe.eu.

ENTSO-E. 2013b. *Interconnected network system grid maps*. Interactive Web tool and map available upon request. https://www.entsoe.eu.

ENTSO-E. 2013c. *Memo 2012*. https://www.entsoe.eu.

Federation of Electric Power Companies of Japan. 2013. *Electricity Review Japan, 2013*. http://www.fepc.or.jp.

GENI. 2013. Africa, Kazakhstan, Saudi Arabia, Jordan, and Nepal power grid maps. Global Energy Network Institute. http://www.geni.org.

Grid Australia. 2013. Map from Australian network operators. http://www.gridaustralia.com.au.

Haubrich, J-H., and W. Fritz 1999. *Study of cross-border electricity transmission tariffs*. Project report to the European Commission, DG XVII/C1.

Hauer, J., T. Overbye, J. Dagle, and S. Widergren. 2002. Advanced transmission technologies. In *National transmission grid study: Issues papers*. Washington, DC: US Department of Energy.

IEA PVPS. 2013. International Energy Agency implementing agreement on photovoltaic power systems. http://www.iea-pvps.org.

International Energy Agency. 2002. *Electricity in India*. Paris: OECD/IEA.

Jaehnert, S., O. Wolfgang, H. Farahmand, and S. Völler. 2014. Transmission expansion planning in Northern Europe in 2030—Methodology and analyses. *Energy Policy* 61:125–139. http://dx.doi.org/10.1016/j.enpol.2013.06.020.

Kanevce, A., I. Mishkovski, and L. Kocarev. 2013. Modelling long-term dynamic evolution of Southeast European power transmission system. *Energy* 57:116–24.

Kessides, I., Z. Bogotic, and L. Maurer. 2007. *Current and forthcoming issues in the South African electricity sector*. World Bank Policy Research working paper 4197. Washington, DC: World Bank.

Kinney, B., P. Crusitti, R. Albert, and V. Latora. 2005. Modelling cascading failures in the North American power grid. *European Physical Journal B* 46:101–7.

Kinyanjui, B., A. Gitau, and M. Mang'oli. 2011. Power development planning models in East Africa. *Strategic Planning for Energy and the Environment* 31 (1): 43–55.

Koç, Y., M. Warnier, R. Kooij, and F. Brazier. 2013. An entropy-based metric to quantify the robustness of power grids against cascading failures. *Safety Science* 59:126–34.

Komoto, K., C. Breyer, E. Cunow, K. Magherbi, D. Faiman, and P. Vieuten, eds. 2013. *Energy from the desert*. Report from the International Energy Agency Implementing Agreement on Photovoltaic Power Systems. Task Group 8. Cambridge, UK: Earthscan/Routledge.

Madrigal, M., and K. Porter. 2013. *Operating and planning electricity grids with variable renewable generation.* World Bank Study 75731. Washington, DC: World Bank.

Madrigal, M., and S. Stoft. 2012. *Transmission expansion for renewable energy scale-up.* World Bank Study 70265. Washington, DC: World Bank.

Masuda, T., H. Yumura, and M. Watanabe. 2008. Recent progress in HTS cable project. *Physica C* 468:2014–17.

Meibom, P., T. Svendsen, and B. Sørensen. 1999. Trading wind in a hydro-dominated power pool system. *Int. J. Sustainable Development* 2:458–83. doi: 10.1504/IJSD.1999.004341.

Meissen, P., and C. Mohammadi. 2010. Bangladesh grid. In *Cross-border interconnections on every continent.* Global Energy Network Institute (GENI). http://www.geni.org.

Messerly, J. 2013. *HVDC Europe annotated.* http://commons.wikimedia.org/wiki/File:HVDC_Europe_annotated.svg; See also http://issuu.com/docs/technology_outlook_2020_lowres, p. 77.

Mexican Department of Energy. 2010. *Prospectiva del Sector Eléctrico 2010–2025.* Mexico City: Secretaria de Enérgia.

Mexican Federal Electricity Commission. 2011. *Programma de Obras e Inversiones del Sector Eléctrico 2011–2025.* Mexico City: Comición Federal de Electricidad.

NERC. 2013. North American Electric Reliability Corporation. http://www.nerc.com.

Niederprüm, M., and M. Pickhardt. 2002. Electricity transmission pricing: The German case. *Atlantic Economic Journal* 30 (2): 138–47.

Nordel. 2008. *The transmission grid in the Nordic countries.* Nordel. (The flow of detailed public information on the Nordic energy systems published by Nordel during previous decades, notably in Annual Reports, has been discontinued or transferred to the larger European organization ENTSO-E.

OECD/IEA. 2005. *Learning from the blackouts.* Energy market experience series. Paris: International Energy Agency. http://www.iea.org/publications/freepublications/publication/Blackouts.pdf.

Pereira, J., G. Parady, and B. Dominguez. 2014. Japan's energy conundrum: Post-Fukushima scenarios from a life cycle perspective. *Energy Policy.* doi: 10.1016/j.enpol.2013.06.131.

Pickard, W. 2014. The limits of HVDC transmission. *Energy Policy.* doi: 10.1016/j.enpol.2013.03.030.

Reinstein, D., A. Mateos, A. Brugman, T. Johnson, and L. Berman. 2011. *Regional power integration: Structural and regulatory challenges.* Central American regional programmatic study for the energy sector. Report 58934-LAC. World Bank Energy Unit of the Latin America and Caribbean Region Sustainable Development Dept. https://openknowledge.worldbank.org/bitstream/handle/10986/2766/589340ESW0P1100toryModule00English0.pdf?sequence=1.

Reuters. 2008. *Indonesia's creaking power grid drags on business.* Feature 27. July. http://www.reuters.com.

Schaber, K., F. Steinke, and T. Hamacher. 2012. Transmission grid extensions for the integration of variable renewable energies in Europe: Who benefits where? *Energy Policy* 43:123–35. doi: 10.1016/j.enpol.2011.12.040.

Schroeder, A., P-Y. Oei, A. Sander, L. Hankel, and L. Laurisch. 2014. The integration of renewable energies into the German transmission grid—A scenario comparison. *Energy Policy* 61:140–50. doi: 10.1016/j.enpol.2013.06.006.

Silverstein, A. 2011. *Transmission 101*. US National Council of Electricity Policy (NCEP) Transmission Technologies Workshop, Denver, CO. http://www.naruc.org.

Sim, K., S. Kim, J. Cho, D. Kim, C. Kim, H. Jang, S. Sohn, and S. Hwang. 2008. DC critical current and AC loss measurement of the 100m 22.9kV/50MVA HTS cable. *Physica C* 468:2018–22. doi: 10.1016/j.physc.2008.05.274.

Sørensen, B. 1981. A combined wind and hydro power system. *Energy Policy* 9 (1): 51–55.

Sørensen, B. 2007. A renewable energy and hydrogen scenario for northern Europe. *Int. J. Energy Research* 32:471–500.

Sørensen, B. 2010. *Renewable energy: Physics, engineering, environmental impacts, economics and planning*. 4th ed. Burlington, MA: Elsevier.

State Grid Corporation of China. 2013. Grid map. http://www.sgcc.com.cn.

Tawney, L., R. Bell, and M. Ziegler. 2011. *High wire act: Electricity transmission and its impact on the renewable energy market*. Washington, DC: World Resources Institute.

Thiel, E. 1951. The power industry in the Soviet Union. *Economic Geography* 27 (2): 107–22.

6

PIPELINE INTERCHANGE

Transport of fuels by pipeline is in widespread use for oil and natural gas, but some biofuel slurry and hydrogen is also moved by pipeline, the latter mainly for industrial use. Figure 6.1 shows a section of a gas pipeline in Alaska. Moving liquid biomass products through pipelines is not different from the current use for oil. If hydrogen gains a prominent place in future energy systems, a need for pipeline transmission will arise, and questions of upgrading natural gas pipelines to transport hydrogen will be asked, considering the increased brittleness of many steel tubes when carrying pressurized hydrogen, as well as the possibly higher leakage rate in the compression units that are usually placed at intervals along the line (Sørensen et al. 2001; Gondal and Sahir 2012; Beaufumé et al. 2013).

It is generally found that after addressing these concerns, the cost of hydrogen pipelines increases to some 25% over that for corresponding natural gas pipelines (Beaufumé et al. 2013). This is also borne out by the estimated cost of a natural gas pipeline at just under 1 million US$ per GW transfer capacity (Gondal and Sahir 2012) and the estimated cost of a hydrogen pipeline cost at just over 1 million US$ (Johnson and Ogden 2012), all for a pipe diameter of 1 meter and a transmission length of 300 km or slightly more. Presently, hydrogen is produced from methane at a cost of around 1 US$/kg, but the cost of production from electricity by electrolysis is up to five times higher (Sørensen 2011). This may not remain true in a future scenario with extended use of hydrogen and electricity production from wind turbines or photovoltaic panels. Such a development could well diminish electrolyzer costs.

The cost estimates for around 1-GW transfer by the pipeline discussed here are similar to those of offshore electric transmission lines discussed in Chapter 5. Because onshore electricity transmission by overhead lines is likely lower by a factor of five, and underground

Figure 6.1   Natural gas pipeline, Alaska (US Department of Transportation, public domain picture).

land-cables are probably also less expensive than offshore ones, the question posed in connection with the Section 5.3 North American scenario of whether one should convert to hydrogen before or after transfer from Canada to the United States or from the United States to Mexico should probably be answered in favor of using electricity for long-distance transmission. This scenario would be even more compelling if the end use is not as hydrogen in the transportation sector, but some other form of energy that would be both cheaper and more efficient to produce from electric power than from hydrogen.

One may speculate whether new pipeline technologies may come online in the future and alter the cost structure. Bolonkin (2010) has suggested transport of natural gas at near-ambient pressure through plastic tubes with diameters of 10 m, elevated into the atmosphere at up to 6 km by the buoyancy of the methane, a scheme that conceivably could also be used for hydrogen. Because methane is a potent greenhouse gas, rupture of the plastic lines could cause serious environmental impacts, but this is not the case for hydrogen, where the lower-than-air density would simply constitute a safety feature compared to underground pipeline transport (Lins and Almeida 2012; Lutostansky et al. 2013). The low cost suggested by Bolonkin for a multigigawatt plastic line floating in the air but moored at intervals will of course have to be further scrutinized, along with pilot experiments.

Charts exist showing the current natural gas pipelines in many parts of the world, similar to the power transmission line maps shown in Chapter 5. An overview is shown in Figure 7.4. However, these are not necessarily relevant for renewable or nuclear energy scenarios that depend on hydrogen as a storage medium as well as a likely energy carrier for the transportation sector, because only very few existing gas pipelines are built to standards that would allow them to be economically modified to carry hydrogen instead of methane. (Issues include steel quality as well as the mentioned pressure and control stations along the pipeline.) Therefore, a hydrogen grid would most likely have to be built from scratch if hydrogen is going to play the role in energy supply that many envisage.

Greenhouse politics should, ideally, rapidly reduce the usage of oil and gas, and well before physical depletion, if scientific warnings are heeded. Coal has an even higher greenhouse warming effect per unit of energy than natural gas or oil, so continued use of coal is seen as demanding carbon removal and safe disposal (Sørensen 2011). CO_2 sequestering after coal combustion is inefficient, and a more workable solution would be to gasify the coal and produce gaseous fuels, of which the cleanest solution is hydrogen production. Because the molecular mass of CO_2 is 3.67 times that of coal itself, the disposal of this residue is a formidable task, which in a scenario where coal plays a significant role cannot be accomplished alone by storage in abandoned gas-extraction fields or mines, but also requires large-scale disposal at sea, in the form of carbonates (see the "clean fossil fuel scenario" in Sørensen [2012]). If coal subjected to these remedies turns out to be economically viable relative to the renewable-energy-based hydrogen, then hydrogen from coal could be a candidate for playing a major role in future energy supply scenarios for the next centuries, particularly for uses in the transportation sector.

The hydrogen employed in several of the scenarios in this book is produced by electric power generated from surplus electricity, which is intermittent and therefore has to be stored before use in the transportation sector or used for regeneration of power. This implies a need for pipeline transport of the hydrogen from the point of production to a hydrogen store and, further, to the hydrogen filling stations for fuel-cell vehicles. The line to the storage facility must be rated at the maximum level of hydrogen production during periods of high electricity surplus, while the lines to the filling stations can be more modestly rated, depending on the size of local hydrogen stores at the filling stations. Regarding the power regeneration part, no additional pipelines are required if the power plants (likely high-temperature fuel-cell installations or gas turbine generators) are the same or are placed in the same location as the hydrogen production facilities, and provided that the highest deficit to be covered by power from hydrogen is no larger than the highest power surplus from intermittent renewable energy. The simulation models monitor the filling of the hydrogen stores as a function of time but assume that the capacity for pipeline transfer is always in place.

Figure 6.2 indicates the pipeline capacity requirement in the reference scenario for Mexico, the one among the Chapter 4 scenarios exhibiting the highest need for regenerating power from stored hydrogen. The time distribution shows highly intermittent production of hydrogen from surplus electricity, reaching a maximum of about twice the total electricity demand (Figure 4.26). The capacity needed for piping hydrogen to power plants is only half as big, corresponding to an entire peak electricity demand. Figure 6.2 illustrates that hours of large surplus and large deficit can follow each other quite closely. The large requirements for pipeline capacity required to perform these operations suggest that hydrogen stores and the hydrogen production/power regeneration plants be sited close together.

Pipeline transport of heat between combined power and heat plants (CPH), or pure district heat plants, and the heat consumers in buildings are plentiful in middle and northern European cities (Figures 6.3 and 6.4). Many installations in large cities with high heat-use density go back to the early twentieth century, and at the end of that century, many small-town areas and suburbs were supplied with district heating based on new, decentralized natural-gas-fired

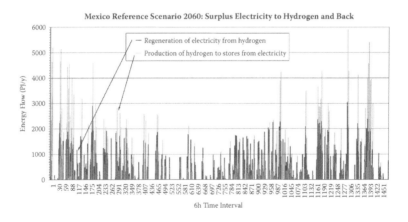

Figure 6.2 Conversion from electricity to hydrogen and from hydrogen to electricity in the Mexican reference scenario of Chapter 4. The numbers are the inputs to the conversion processes, before conversion losses. The maximum regeneration level thus indicates the hydrogen pipeline capacity required, while the pipeline capacity for produced hydrogen is the electric input times the conversion efficiency of the electrolyzers, assumed to be 0.89.

Figure 6.3 District heating pipes ready to be laid down (Denmark, photo by author).

CPH plants. In scenarios with more efficient heating systems, many areas of the world will have much less demand for space heating than in the twentieth century, but the demand for hot water remains. The scenarios considered in this book satisfy part of the heat demands by heat pumps and part by solar building-integrated collectors. Any heat demand remaining will be served by waste heat from power plants (now likely fuel-cell based), but these plants may be placed either centralized or decentralized (Sørensen 2010). If they are placed centrally, there will be a need for heat transmission, possibly through already

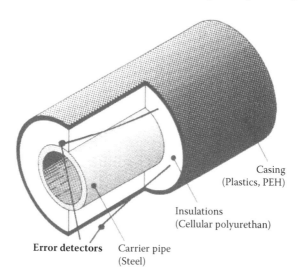

Casing
(Plastics, PEH)

Insulations
(Cellular polyurethan)

Error detectors Carrier pipe
(Steel)

Figure 6.4 Design of typical Danish district heating pipe (Danish Energy Agency, 1993; used with permission).

existing lines for district heating. In the case of decentralized solar collectors (combined PVT panels, cf. Figure 4.6), the stores being part of any solar heating system could be placed centrally or at least be shared by several buildings because of the reduced heat losses through the store surface, relative to building-located stores. This would again cause a need for district heating lines, but only between the houses with solar panels and the communal stores. The cost of heat lines is greatly reduced if they are laid down (at depths of typically a meter) during urban development, rather that in existing towns with solid pavements and several other lines crisscrossing the underground (e.g., electricity, telecommunication, and gas lines).

References

Beaufumé, S., F. Grüger, T. Grube, D. Krieg, J. Linssen, M. Weber, J. F. Hake, and D. Stolten. 2013. GIS-based scenario calculation for a nation-wide German hydrogen pipeline infrastructure. *Int. J. Hydrogen Energy* 38:3813–29.

Bolonkin, A. 2010. Aerial high altitude gas pipeline. *J. Natural Gas Science & Engineering* 2:114–21.

Danish Energy Agency. 1993. *District heating in Denmark*. Copenhagen.

Gondal, I., and M. Sahir. 2012. Prospects of natural gas pipeline infrastructure in hydrogen transportation. *Int. J. Energy Research* 36:1338–45.

Lins, P., and A. Almeida. 2012. Multidimensional risk analysis of hydrogen pipelines. *Int. J. Hydrogen Energy* 37:13545–54.

Lutostansky, E., L. Creitz, S. Jung, J. Schork, D. Worthington, and Y. Xu. 2013. Modeling of underground hydrogen pipelines. *Process Safety Progress* 32 (2): 212–16.

Sørensen, B. 2010. *Renewable energy*. 4th ed. Burlington, MA: Elsevier.

Sørensen, B. 2011. *Life-cycle analysis of energy systems: From methodology to applications*. Cambridge, UK: RCS Publishing.

Sørensen, B. 2012. *Fuel cells and hydrogen*. 2nd ed. Burlington, MA: Elsevier.

Sørensen, B., A. Petersen, C. Juhl, T. Pedersen, H. Ravn, C. Søndergren, P. Simonsen, et al. 2001. Scenarier for samlet udnyttelse af brint som energ-ibærer I Danmarks fremtidige energisystem. *Texts from IMFUFA*, No. 390, Roskilde University. http://rudar.ruc.dk/handle/1800/3500, filename IMFUFA_390.pdf. Summary published in 2004, *Int. J. Hydrogen Energy* 29:23–32.

7

OTHER TRADE OPTIONS

There is a long tradition of moving solid and liquid fuels by ship, train, or truck. The energy expenditure for moving one ton of oil on a highway is at best 1.25 MJ/km (using a large train-type multiunit motortruck of the kind you meet when you travel on the strictly linear highway from Adelaide to Darwin in Australia), more typically 2–4.25 MJ/km (average heavy trucks), and several times higher if light trucks are used. All of these values are higher than hauling by train, the energy cost of which is about 0.2 MJ/km, and much higher than transport on fuel tankers at sea, with energy costs below 0.05 MJ/km (US DoE 2005; Pootakham and Kumar 2010; Sørensen 2012, chap. 12). Figure 7.1 shows a typical contemporary tanker vessel.

For comparison, the energy needed to pump bio-oil (biomass crudely liquefied by pyrolysis) through a 100-km pipeline of diameter 5 cm is 3.3 MJ/km (Pootakham and Kumar 2010). Hamelinck, Suurs, and Faaij (2005) discussed the cost of delivering raw or transformed biomass to the Netherlands and found that delivery from Latin America was cheaper than delivery from Sweden or Finland, both as regards transport and in choosing the location for the liquid-fuel conversion process. The transport in this case consists of long-distance ship hauls plus varying distances of land transport by truck, reflecting the location of forest or other biomass sources in the two mentioned regions, both of which are known for abundant availability of bioresources. A problem with this type of approach is that it is strictly valid only due to the prevailing conditions, where international ship transport is not taxed, and also where, if a major export route were to be established between Latin America and Europe, one might expect prices to rise to eventually match that of European alternatives rather than continue to take advantage of the lower production costs in South America. (Indeed, a similar scenario has happened with electricity export from

Figure 7.1   Oil tanker at sea (*Mærsk Nautilus*) (photo by Mærsk Tankers, used with permission).

Norway to countries farther south in Europe, with the prevailing costs now reflecting market prices in central Europe rather than production costs in Norway.) The transport energy costs for solid or liquid biofuels assumed by Hamelinck, Suurs, and Faaij (2005) agree with those quoted here for large ships, are three times higher than those quoted for train transport, and are four times lower than those given for road transport, possibly reflecting particular data sources (e.g., using Swedish data for train transport) or market situations. Still, the low values for truck transport do not appear representative.

The total cost of transporting and handling solid biomass from Latin America to a European destination is estimated as about 40 euro per dry ton, to be added to an assumed biomass cost of around 25 euros per dry ton for energy crops (i.e., sources such as wood rather than sustainable use of biomass residues) grown in Latin America. The direct ocean-tanker shipping cost of about 8 out of the 40 euros per ton quoted previously had gone up to around 15 euro per ton by mid-2005 for oil shipment from the Middle East to the United States (Orwel 2005). Since then, shipping rates have fluctuated substantially under the influence of the ongoing global economic crisis.

Transportation of hydrogen factors into many recent studies looking at scenarios with increasing penetration of hydrogen as a useful fuel in the transportation sector. The studies generally find that pipeline transfer is the most economic solution for larger hydrogen quantities (350–800 US$/ton for a 50–300 km haul, the assumed energy flow is 0.14 GW, cf. Chapter 6), but in land areas such as the United States, pipelines would increasingly be competing with liquefied hydrogen truck transport and being cheaper only for very large gas flows. This all depends on the price of hydrogen, as there are more substantial energy losses associated with liquefaction and storage of liquid hydrogen than losses occurring in pipelines. For small hydrogen demands and short distances, road transport of hydrogen as a compressed gas is advantageous, at an estimated cost of 900 US$/ton for a 50 km haul (Yang and Ogden 2007). The energy content of a ton of hydrogen is some three times higher than in a ton of oil, but still the delivery costs quoted are higher than pure long-distance transmission costs. One reason is that they include the cost of the associated hydrogen fueling stations, the design of which depends on the mode of delivery. The pipeline transmission cost per kilometer is found to be diminishing logarithmically as the distance increases. The higher cost relative to endpoint delivery pipelines is due to the higher cost of laying down pipelines in urban or semi-urban areas.

The total amounts of oil and natural gas currently shipped internationally are large, due to the uneven distribution of resources as compared to the location of major load areas. Current annual shipments of crude oil and oil products by tankers is about 5.6×10^{15} kg (ITOPF 2013). Figure 7.2 shows the main oil shipping itineraries. Figure 7.3 details the amounts shipped and also indicates overland shipping by means other than marine. The similar data for natural gas shipments are shown in Figure 7.4, here distinguishing between liquefied (LNG) transport and pipeline transport.

The scenarios in this book employ biofuels, particularly in the transportation sector. This may entail an amount of intermittency due to concentration of harvest times during particular periods of the year. In some cases, the harvested biomass can be stored at the point of origin or at the plant responsible for the subsequent transformation into demanded biofuels. In any case, there will be a need for transportation of the biomass, which in these scenarios consists entirely of residues from primary products such as grain or vegetables or hardwood. This

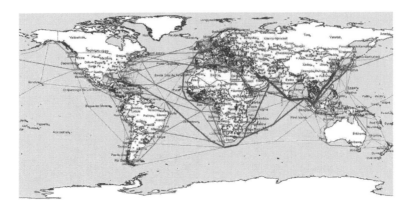

Figure 7.2 Major oil shipping routes 2011. The heaviest lines indicate an annual cargo hauling of over 300 million tons (ITOPF [2013]; used with permission).

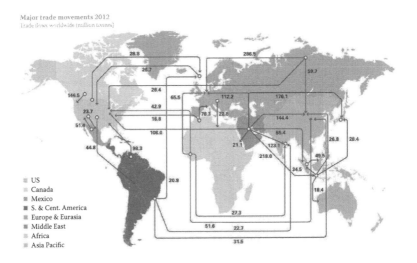

Figure 7.3 Major oil shipments 2012, chiefly by sea (million tons) (BP [2013]; used with permission).

can be expensive due to the low-volume density of certain biomass residues such as straw and loose leaves or twigs. In many cases, a solution will be to compress these materials before transport (while counting the energy used in doing so). Farm compressors and pelletizers are in use today to do just that. Because pellets and other biomass resources are today used extensively for plain combustion in power plants, the necessary transportation equipment is already available, and the transport is considered economically viable for typical distances between farms, forests, or other growing areas and the location of power or heat plants.

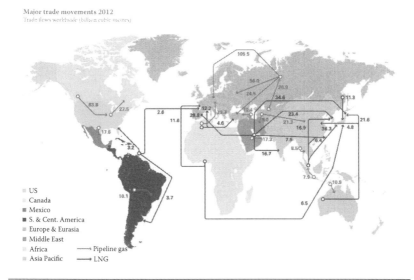

Major trade movements 2012
Trade flows worldwide (billion cubic metres)

US
Canada
Mexico
S. & Cent. America
Europe & Eurasia
Middle East
Africa ⟶ Pipeline gas
Asia Pacific ⟶ LNG

Figure 7.4 Major natural gas shipments 2012, chiefly by pipeline and by LNG carrier (10^9 m^3) (BP [2013]; used with permission).

The same is not the case for dealing with the secondary residues arising after converting the residues to biofuels. It is essential that nutrients be returned to the soil to ensure long-term sustainability or near sustainability (considering that 100% recycling of nutrients is never possible). The recycling is fairly easy for biogas production, where the residue is already in the form of a fertilizer, and even one of higher quality than that of the original biomass feedstock. Still, only a few biogas plants currently recycle the fertilizer, claiming that the cost of transportation back to farms is too high. This economic argument is stronger for the types of biofuel production residues arising from conversion into ethanol, methanol, biodiesels, and so on, because the energy extraction leaves the nutrients trapped in more complex residues. For example, following the high-temperature gasification step involved in several liquid biofuel schemes, the residue is a tar or charred solid, which would have to be subsequently processed to be useful as a fertilizer on the fields.

In any case, the cost of making the biofuel residues suitable for nutrient recycling and then actually transporting them back to agriculture or silviculture is high, and this cost must be included in the price of the biofuels to make a fair full-cost comparison with other ways of satisfying transportation energy needs, e.g., by hydrogen. Just as environmentalists are criticizing the use of edible biomass for fuel production (first-generation biofuels), one would have to question the route from residues to biofuels (second-generation biofuels) if nutrient recycling is not part of the scheme.

The attitude behind the scenario construction in this book is that if the hydrogen/fuel-cell alternative encounters technical problems, then the transportation sector has only battery-operated vehicles and biofuel vehicles left as currently identified options. In this case, it may well be that the use of biofuels with nutrient recycling is less expensive than the use of batteries, while in other cases, there may not be enough electricity production to cover the entire transportation sector by use of electric vehicles when one considers the demands for dedicated electricity, industrial process heat, and possibly heating by heat pumps.

References

BP. 2013. *BP statistical review of world energy June 2013*. London: BP p.l.c. http://www.bp.com/statisticalreview.

Hamelinck, C., R. Suurs, and A. Faaij. 2005. International bioenergy transport costs and energy balance. *Biomass & Bioenergy* 29:114–34.

ITOPF. 2013. WebGIS Map. International Tanker Owners Pollution Federation Limited. http://www.itopf.com/information-services/data-and-statistics/gis-map.

Orwel, G. 2005. Tanker rates rise on higher US shipments. *Oil Daily*, July 14, 1.

Pootakham, T., and A. Kumar. 2010. A comparison of pipeline versus truck transport of bio-oil. *Bioresource Technology* 101:414–21.

Sørensen, B. 2012. *A history of energy*. New York: Earthscan/Routledge.

US DoE. 2005. *Transportation energy data book*. 24th ed. Washington, DC: Oak Ridge National Laboratory and US Department of Energy.

Yang, C., and J. Ogden. 2007. Determining the lowest-cost hydrogen delivery mode. *Int. J. Hydrogen Energy* 32:268–86.

PART II
ENERGY
STORAGE

Storage of energy is one obvious way of handling variations and inter-mittency. Fuels have a built-in storage option, so to speak. At least this is true for the fuels that have been in use for the longest time, wood and coal, which can simply be stored on the ground, preferably out of the rain, but it is also largely true for other fossil fuels such as oil (stored in simple vessels and containers) or natural gas (stored in closed containers, possibly under elevated pressure). Also, hydropower plants that are based on elevated reservoirs have the storage option integrated, as the water can simply be held back until demand arises. These integrated solutions, where energy source and storage option are closely related, are discussed in Chapter 8.

Chapter 9 deals with some other storage options, characterized by requiring a more intricate, separate device for accomplishing the storage and retrieval of energy, a device that may not be connected with or located close to the energy source or production equipment. Examples of centralized stores of this kind are hydrogen or compressed-air caverns, communal heat stores, as well as flywheels and batteries. The latter two categories can be used at many different scales, and Chapter 10 focuses on storage applications in highly decentralized settings, such as building-integrated stores or stores in portable equipment.

While the energy transmission exchange and other trade options discussed in Part I are limited by the availability of surplus energy in one place to export at the particular time when it is requested in another place, the energy storage options can be used in all cases without such limitations, provided that one can afford to build stores of sufficient capacity.

8

SYSTEM-INTEGRATED STORAGE

While some hydro installations only make use of the run of a stream of water, power plants in several of the most mountainous regions can benefit from natural or manmade water reservoirs, allowing more-or-less complete timewise decoupling between filling the reservoirs by precipitation and letting the water down through the turbines to satisfy an energy demand. Figure 8.1 shows such a reservoir. Figure 8.2 shows the yearly variations of the combined water levels of all the Norwegian reservoirs used for hydropower generation. In addition to the results for a typical year, the figure also provides curves for the highest and lowest water-level years in the recent past. They primarily reflect the variations in annual precipitation, but are also influenced by variables such as temperature and timing of snowmelt that affect the ratio between runoff and evaporation. The high-latitude location of Norway makes the runoff available for hydropower exploitation highly dependent on when the snow and ice accumulated during winter melt, usually sometime in May and June (cf. Figure 3.3). From August, the hydropower utilization exceeds the filling of the reservoirs, and the water level drops more or less regularly until the following May. The water-level difference between high and low for 2005 is equivalent to a potential power production change of some 50 TWh, leaving much room for using the hydropower to smoothe intermittent energy production attached to the same grid system. For example, in the scenario depicted in Figure 5.21, Norwegian hydro covers the entire intermittency deficit of a 100% Danish wind-based electricity system by 5 GW of transmission capacity between Norway and Denmark, and with an influence on hydro reservoir filling amounting to less than or about 2% (Sørensen 1981). Figure 8.3 further illustrates the issue by giving the weekly energy input into the Norwegian

Figure 8.1   Manmade hydro reservoir in Tasmania (video still by author).

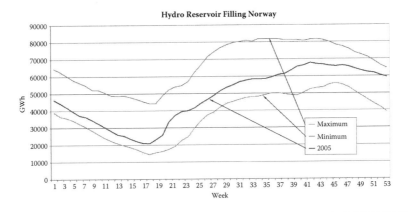

Figure 8.2   Filling of Norwegian hydro reservoirs through the year 2005, as well as for years with maximum and minimum filling recorded, expressed as energy content when used to generate hydropower (based on data from NORDEL [2005]).

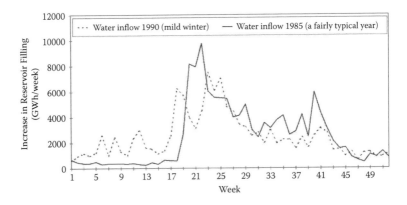

Figure 8.3   Weekly water influx into Norwegian hydro reservoirs through the years 1985 and 1990, expressed in energy units (from Meibom, Svendsen, and Sørensen [1999]; used with permission).

reservoirs during both a typical year (1985) and an atypical year (1990), rather than the accumulated energy level of Figure 8.2 (with water level being expressed in stored energy units).

Hydro reservoirs are sometimes used to store water pumped upwards from a lower reservoir at times of surplus power production from some type of non-hydrogenerators (a solar PV example is worked out in Margeta and Glasnovic [2012]). In most cases, this view of hydro reservoirs as a plain in–out store is not required, because the joint operation of hydropower and intermittent power production units in the way demonstrated in the previous Norway–Denmark example, holding back water when other sources produce a surplus, is often possible and cheaper than including upwards pumping. The pumped hydro stores in operation cater to a different situation of, say, shaving a daily repeating peak by letting the turbine from the store cover the peak by using the water pumped upwards during the preceding night.

In geographical settings where topological height differences are insufficient, one can consider establishing artificial underground cavities to serve as the lower reservoir for a pumped hydro system, with the upper reservoir being at ground level (lake or sea). Clearly, this solution is considerably more expensive than schemes taking advantage of naturally occurring potential repositories of water at elevated levels (Inage 2009).

Fuels are normally inherently storable, most easily for solid fuels, quite easy for liquid fuels, and also possible for gaseous fuels, but requiring consideration of the bulkiness of fuels with low-volume energy density such as hydrogen. A survey of storage options for hydrogen may be found in the work of Sørensen (2012). The new biofuels envisaged in many energy scenarios for the near future can essentially be stored in the same was as fossil fuels, in containers for the liquid variants and compressed or liquefied for the biofuels to be used in gaseous form. The conventional solid biofuels (wood, straw, shavings, tiles, and briquettes) that have played a major role through human history should, for health reasons, only be combusted under strict environmental control (e.g., by flue-gas scrubbers and electrostatic filters aimed at diminishing particulate emissions in visible and invisible size ranges, volatile organic components, etc.), which will likely be possible only for large-scale use (such as in power plants) and

Figure 8.4   In some cases, it is possible to store biomass directly. For cereal crops, grain and straw may be separated and the straw compressed for transport and storage before further conversion, notably to biofuels (photo by author).

not for decentralized applications (wood stoves and other domestic fireplaces). Storage of solid biomass has traditionally taken place both before and after processing or treatment (Figures 8.4 and 8.5). However, energy uses of food grain and construction hardwood are not acceptable in a sustainable scenario, so only storage of residues such as compressed straw, used wood scraps, twigs, and other residues collected from the forest floor as part of ecological forestry maintenance are of interest.

An expanded use of new biofuels (rather than hydrogen) in the transportation sector will fairly certainly lead to environmental problems similar to those from the current oil-based fuels, except for the CO_2 emission, which will be compensated for by the plant's assimilation of CO_2 during growth, at least in the case where the biofuels are based on short-rotation plants (such as cereal crop residues). Burning of wood or wood-based biofuels (e.g., methanol, except when this is produced from sustainably managed forestry residues and not from timber) is not CO_2-neutral, because the assimilation took place 50–100 years ago when, as mentioned previously, the

Figure 8.5   Wood has traditionally been cut into suitable pieces and stored for later combustion (photo by author). Direct combustion is considered less environmentally viable than combustion of biofuel products, at least in the case of small-scale woodstove combustion.

atmosphere–ocean system was capable of handling the fluctuations in CO_2 levels, and thus, the current emission from wood combustion is not directly being compensated for.

Vehicles using biofuels would require emission control similar in nature to that of current vehicles with diesel engines, aimed primarily at NO_x and particles, which are fewer than those from burning solid biofuels but still in significant numbers. Presently employed methods of controlling such emissions are decreasingly efficient for smaller devices, and it is therefore suggested that biofuel combustion in passenger vehicles should only be a transitional solution, eventually to be replaced by hydrogen fuel cells or battery-operated electric motors, presumably in a hybrid mode (Sørensen 2012).

References

Inage, S. I. 2009. *Prospects for large-scale energy storage in decarbonised power grids.* Working paper, International Energy Agency, Paris.

Margeta, J., and Z. Glasnovic. 2012. Theoretical settings of photovoltaic-hydro energy system for sustainable energy production. *Solar Energy* 86:972–82.

Meibom, P., T. Svendsen, and B. Sørensen. 1999. Trading wind in a hydro-dominated power pool system. *Int. J. Sustainable Energy* 2:458–83. doi: 10.1504/IJSD.1999.004341.

NORDEL. 2005. Spreadsheet data report No. 2005-00537-01, a supplement to *Annual report 2005, No. 2005-00336-01*. (Downloaded in 2006 from a now-defunct Nordic Power Utilities Association website.)

Sørensen, B. 1981. A combined wind and hydro power system. *Energy Policy* 9:51–55.

Sørensen, B. 2012. *Fuel cells and hydrogen*. 2nd ed. Burlington, MA: Elsevier.

9

STORAGE IN
DEDICATED FACILITIES

While the storage options naturally connected to certain types of energy installations are usually of modest cost, this is rarely the case for dedicated stores built solely to take care of mismatch between production and demand. This is one reason why many competing solutions have been explored and why the different paths of development have created a shifting environment, where one storage solution is preferred at a given time, while a different one is preferred at another point in time. The following discussion of several such storage technologies is divided into stores for low-quality energy (heat at low temperatures, Section 9.1) and stores for high-quality energy (electric power, mechanical energy, and chemical energy, Section 9.2).

9.1 Storage of Low-Quality Energy

Low-quality energy systems typically use heat at temperatures up to about 100°C in the form of flows of water or air, and somewhat higher for storage media suitable for that. Examples of technologies requiring energy storage are solar heating systems, which exhibit particular intermittency problems: There is no solar input during night, and those areas requiring space heating usually have very little solar radiation available during the coldest season. The day-to-night supply–demand mismatch can be taken care of by a modest-size water container, no matter whether the heat distribution system uses water or air (in the latter case, a heat exchanger is part of the system). Such solar systems, sketched in Figure 9.1, are in use in many regions of the world, providing hot water for the inhabitants of a building during summer, with extensions into spring and autumn, depending on system size and the radiation resources available at the latitude where the system is operating. More difficult to cope with are the systems aimed at providing

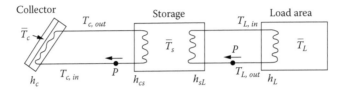

Figure 9.1 Design of typical solar heating system for a building, including circulation of a fluid such as water from the heat store (also normally water) to the collector and a second circuit between store and load areas, where the choice of using water or air is often settled by the presence of an existing setup. The several temperature levels involved determine the efficiency of the solar energy conversion (Sørensen 2007; used with permission).

space heating year-round. Here, seasonal storage is required, which is extremely difficult to achieve by liquid water storage in a detached building, at least at currently prevailing insulation levels (which determine the amounts of energy needed for space heating, typically peaking during cold winter periods). In consequence, communal storage is often invoked for high-latitude locations, taking advantage of the declining surface-to-volume ratio (and thus the relative heat loss through the surface) with increasing storage size. Typical communal heat stores are in underground reservoirs or insulated lakes featuring some type of removable insulating cover (Sørensen 2010).

Seasonal solar storage systems may include a heat pump to raise the temperature of the stored medium or the part of it flowing to the rooms and hot-water outlets of the buildings being served in cases where the flow falls short of the minimum inlet temperature for (various parts of) the building heat-distribution system. Because the efficiency of a thermal solar collector drops with increasing water inlet temperature, one could also use a heat pump to lower the inlet temperature of the fluid passed through the solar collectors. These are subtle design issues depending on the overall size of the store relative to the load, and several simulations of various systems, plus measurements on a few systems actually built, have been performed.

Needless to say, there are many materials other than water that may be used for storage (metals, rock, gravel), including some for which heat is used to induce reversible chemical reactions and phase-change materials such as hydrated salts that have loss mechanisms different from that of plain uses of heat capacity, but also a different cost structure. Such solutions have been discussed in several review articles (Sharma et al. 2009; Pinel et al. 2011; Li and Wang 2012;

Tatsidjodoung, Pierrès, and Luo 2013), but few practical solutions have emerged, and the recent reviews do not differ substantially from early ones (such as Sørensen [1979, Section 5.2], repeated in subsequent editions).

Solar hot-water systems used in low-latitude geographical areas with year-round adequate average solar radiation only need to store energy collected during the daytime for load coverage during periods without radiation, a task carried out by a small hot-water store (on the order of one cubic meter) for a family dwelling. In rare special circumstances where such a hot-water volume is inconvenient, a higher heat-capacity material may be used instead (Haillot et al. 2012). The challenge for solar heat systems aimed at more general heat supply is to cope with seasonal variations. As mentioned previously, in many situations this calls for communal solutions if plain heat-capacity stores are employed.

Before looking at energy system solutions in particular geographical settings, an overview of the variations in space-heating requirements with locations may be instructive. Figure 9.2 shows the energy requirement for space heating at critical times of the year as calculated for a 2050 scenario (Sørensen and Meibom 2000) assuming the UN middle projection on population and making a (hopeful) assumption that all world citizens have their basic energy requirements (here for heated shelter) satisfied using modern technology, which for space conditioning means dwellings built with proper insulation and architecture minimizing artificial heating and cooling needs. Sørensen (2008a) discusses policies of increasing aggressiveness in achieving low-energy input for creating acceptable human indoor environments and considers, for space heating, three levels of average energy spending defined by a requirement of 36, 24, and 18 W per capita per degree Celsius to raise the indoor temperature for rooms to 16°C in dwellings and workplaces with a total floor area of 60 m² per capita, assuming that indoor equipment and activities will provide the further rise to 20°C. The three technology levels are argued to roughly correspond to the best technology available by 1980, by 2010, and by 2050, but even for 2050 using only technology already available and used today in buildings with the highest energy standard. The average energy requirements depicted in Figure 9.2 use the least ambitious scenario assumption of 36 W per capita per °C, which, however, is

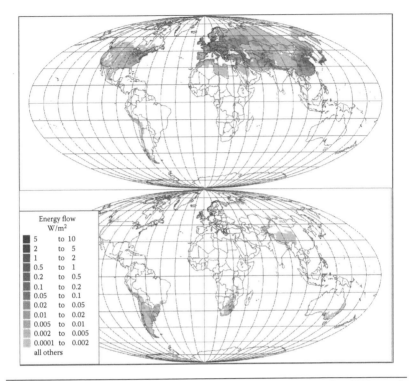

Energy flow
W/m²
5   to 10
2   to 5
1   to 2
0.5   to 1
0.2   to 0.5
0.1   to 0.2
0.05   to 0.1
0.02   to 0.05
0.01   to 0.02
0.005   to 0.01
0.002   to 0.005
0.0001 to 0.002
all others

Figure 9.2   Space heating demand (W/m²) in January (above) and July (below) for a 2050 scenario with high fulfillment of human needs and use of proven efficient energy technologies (here build-ing design), which leads to large areas not requiring any artificial energy input for space heating. Because population size per square meter is folded into the demand graph, cold regions with low population have lower demand than less-cold regions with higher population density (Sørensen and Meibom 2000; used with permission).

still considerably better than the current average technologies used in any globally selected region.

Considering a single detached family dwelling with a heat load, providing seasonal storage of heat from solar heating of water is not a good choice (as noted previously), because conventional solar heat-ing systems use the water from the store to pass through the col-lectors and, therefore, will have a low efficiency when the storage temperature approaches 100°C, where the system will be unable to store more heat before the load has lowered the store temperature. Considerations of overly high or low inlet temperature to the solar collectors restrict the amount of energy stored to correspond to a time period that is lower than required for seasonal displacement. It makes no difference whether heat exchangers are used to allow a different fluid or gas to be employed in the collector circuit. As mentioned previously, making the store larger will lower its temperature to a level that is inadequate for space heating during winter, even with air circulation (which has a lower temperature demand than water radiators), and would still definitely require a heat exchanger. Although a heat exchanger may be used, this is not absolutely needed for water circulation, as one may circulate the same water through the solar collector and the building radiators.

A remedy for the temperature conundrum of solar thermal systems is to add a heat pump that can raise the temperature, when it is too low, to that required by the heat distribution system, but this will entail the use of electricity to drive the compressor of the heat pump (Sørensen 1979). Figure 9.3 is derived from a model study of such a heat pump system for a location in Turkey (Yumrutaş and Ünsal 2012). They consider a 100 m² poorly insulated house with a huge heating demand stated to be 10 kW (presumably the maximum value) and place the spherical water storage tank underground, investigating the heat buildup for different soil surroundings (clay/gravel, limestone, granite). It takes a couple of years of soil gradient buildup for

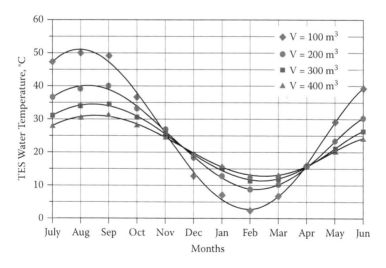

Figure 9.3 Temperature variations from modeling an underground thermal energy store (TES) associated with a detached-dwelling solar collector (20 m²) and with a heat pump system used to raise the temperature of heat drawn from the store when needed as a function of storage volume (V). The system is assumed to be installed at 37° N in Turkey (Yumruta and Ünsal 2012; used with permission).

the system to exhibit the same seasonal variations each year. Various geometries of underground stores have been proposed and studied for many years, with one early work being that of Brüel, Schiøler, and Jensen (1976).

A similar modeling of a single-family dwelling with solar collectors (40 m^2 and up) and underground storage in a sand bed, but without a heat pump, has been made for a location in Virginia (Sweet and McLeskey 2012). In the absence of a heat pump to boost the temperature, they find that a moderately small storage bed of 15 m^3 is optimal. However, in both the Turkish and the US cases, the buildings modeled are so poorly insulated that equivalent energy savings could be accomplished by a simple investment in the building shell, and at a much lower cost than that of the system with solar energy plus storage. It is generally amazing that so many studies focus on fancy energy-supply systems while neglecting the economically much more attractive efforts to improve the efficiency on the demand side by building or retrofitting the house to the best (or at least to a better) energy standard. The criterion should be that the additional investment cost should be lower than the cost of the energy saved over the operating lifetime of the house, ideally taking into account expected price increases of energy over the lifetime of the building, typically 50 to over 100 years and higher in Europe than in North America.

Moving on to communal thermal stores, a number of German projects have explored seasonal stores for buildings equipped with solar thermal panels based on hot water, gravel/water beds, aquifers, and vertical borehole concepts, with some of these combined with heat pumps (Bauer et al. 2010; Marx et al. 2011). Figure 9.4 gives some operating results for a borehole system serving 300 apartments in Neckarsulm, a setup similar to the one proposed by Brüel, Schiøler, and Jensen (1976). It took 5 years to heat up the underground soils to a stable cycling state, and the 5670 m^2 of solar collectors now satisfy some 45% of the heating loads. Projects like this, where district heating lines carry the heat between the buildings and stores involved, can easily incorporate other forms of energy supply if—as in the German case—the solar system is unable to cover the heat load at all times of the year. However, if sustainability is the aim, the auxiliary energy systems should also be renewable. Some solar seasonal storage systems

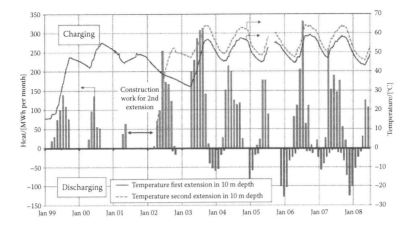

Figure 9.4 Performance of a seasonal borehole thermal store employed in a German solar heating project at Neckarsulm (Bauer et al. 2010; used with permission). For a sequence of winter days, daytime solar panel charging and nighttime discharge to heat loads are displayed in terms of heat-transfer bars (left scale), while store temperatures are shown as a curve (right scale). The second curve appearing three days into the sequence reflects an increase in borehole depth carried out on January 2.

capable of covering annual building heat loads without assistance have been modeled by Sørensen (2010) using simulation techniques.

9.2 Storage of High-Quality Energy

Hydro reservoir stores are the most extensively used form of storing high-quality energy, as discussed in Chapters 5 and 8. Hydro reservoirs are used either as part of power-exchange arrangements between utility systems or within a single electricity system (Figure 2.3), with or without pumping water upwards into the reservoirs. Competing technologies for electricity storage and regeneration include underground compressed-air stores, flywheel and battery stores, and storage in the form of hydrogen in underground caverns, in local containers, in chemical energy stores such as metal or complex hydrides, or as cryo-adsorbed gas in carbon materials (Sørensen 2010, 2012a). Other recent electricity-relevant storage reviews are presented in works by Chen et al. (2009), Inage (2009), and Díaz-Gonzáles et al. (2012).

During the period of transition from depletable to renewable energy sources, possibly extending over several future decades, an obvious alternative to energy storage for handling intermittency is to use fuel-based backup systems, such as the ones now existing in most regions. Some decades ago, this was made difficult by the extended start-up times for some fuel-based power plants, notably coal-fired or nuclear plants. The solution was to have a surplus of power plants online as "spinning reserves," i.e., plants operated at part load so that their power could be turned up or down within certain limits at short notice. For example, such swift availability is made easier by using pulverized coal in coal-fired plants, allowing much faster regulation than with solid coal. Figure 9.5 compares some start-up or regulation times for various technologies currently employed in the Japanese power system. It is seen that even today, no application of fuel-based backup systems can act as swiftly as reservoir-based or pumped hydro. Natural-gas turbines or pulverized-coal-fired plants can deal with variations in uncontrollable components in the energy system, except for those of the shortest duration. As already suggested, such variations will be rare for renewable energy systems because of the dispersed nature of solar or wind energy systems, making it unlikely that the wind should come

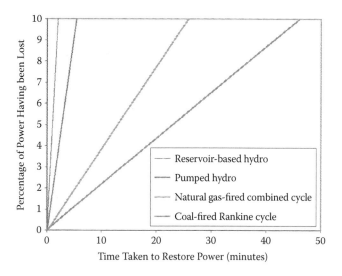

Figure 9.5   Typical times (abscissa) needed for various sources of stored energy or backup power to come online and compensate for the fallout of primary power due to intermittency or failure. The ordinate is the percentage of overall power falling out (based on data from Inage [2009] and Japanese Wind Power Association [2004]). The delay in coming online and beginning to build up compensating power ranges from 1 second for reservoir-based hydro to 6 seconds for pulverized-coal-fired power stations.

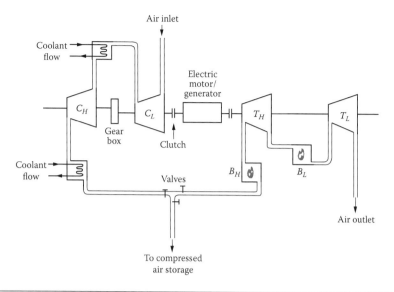

Figure 9.6 Sketch of the layout of a German compressed-air storage facility at a power plant at Huntorf, indicating compressors (C), burners (B), and turbines (T) (Sørensen 2007; used with permission).

to a stop for all turbines at once, or that clouds should suddenly deprive all solar panels of radiative inputs.

Energy storage underground in natural or manmade compressed-air reservoirs is today used in a few places, notably for providing peak power, i.e., chiefly diurnal storage (Figure 9.6). The losses of this storage cycle may exceed 50%, comprising energy for compression and heat for reentering the air into a gas turbine for regeneration of power, plus the turbine's thermodynamic loss. Countries using natural gas often see a need to store a sufficient amount of gas to safeguard against supply disruptions caused by breakage of gas pipelines (many of which are offshore and could thus be cut by fishing trawlers or the passage of submarines) or an interruption in shipped supplies of liquefied gas due to war or political boycotts. European countries importing Russian natural gas by pipeline are concerned about politically motivated disruption of supply. Underground storage of natural gas in reservoirs has thus become fairly widespread. Such reservoirs are usually excavated in salt domes (upward extrusions of salt deposits typically found in soil areas formed by meltwater river deposits during ice-sheet retraction) by water flushing, possibly followed by lining to reduce losses, or in aquifer bends, where nearly impermeable layers such as clay are

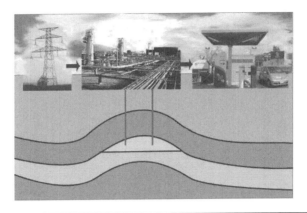

Figure 9.7 Aquifer underground store used at Stenlille, Denmark, for storage of natural gas (Sørensen 2012a; used with permission).

present above and below the aquifer, and where an upwards bend allows gas to be stored locally without flowing away with the water in the aquifer (Figure 9.7). Both of these reservoir types, available in some geological formations in both northern Europe and North America, are considerably cheaper to establish than forming reservoirs by crushing the rock material prevailing at other geological locations. With modest expenses for sealing (in the salt dome case), the natural gas pressure can be maintained at about 20 MPa with modest gas leakage.

In future uses for hydrogen storage, the same reservoirs can be operated with similarly small leakage rates at hydrogen pressures of some 5 MPa, determined by the relative size difference between hydrogen and methane molecules (Sørensen 2010). These reservoir types can provide seasonal storage and thus help to cope with any supply–demand mismatch in fuel-based energy systems, as well as in dealing with future intermittency problems posed by variable energy resource flows, in cases where the renewable energy surpluses can be transformed into hydrogen. The later regeneration of electricity may use gas turbine technology or the higher-efficiency but also higher-cost fuel-cell technologies. Several scenarios demonstrating use of these options on a national or international power supply system are discussed in Sørensen (2008b), and geographically more limited scenarios making use of energy storage to obtain large penetration of renewable energy sources can be found in the work of Glasnovic and Margeta (2011), Tsuchiya (2012), and Elliston, Diesendorf, and MacGill (2012). Scenarios considering only short-term storage are abundant, but they cannot admit large quantities of intermittent resources and therefore tend to fall back on continued use of depletable resources. (For the situation in Japan, see the works by Fujino [2008] and Zhang et al. [2012a] and the more general scenario in [2012b].)

In the North American scenario discussed in Chapter 4, Mexico has a very modest hydrogen storage capacity capable of holding 20 hours of power demand. Figure 4.30 gives the variations in filling during the year. The fact that such modest requirements suffice for handling intermittency of the renewable energy sources is due to the substantial power transmission lines assumed to be in place between the North American countries. Such transmission lines should be capable of handling most mismatches (see Figures 5.23–5.25) with the help of substantial hydro storage reservoirs in the United States and particularly in Canada.

For the US part of the North American reference scenario, hydrogen stores totaling an energy-holding capacity of 800 EJh/y (some 10 days of transportation-sector demand for hydrogen) are considered to be in place, and the predicted exploitation of these stores over the year is displayed in Figure 9.8. It is seen that the intermittency of the power sources producing the hydrogen (solar and wind) is fairly evenly distributed over the year, with most solar energy during summer and

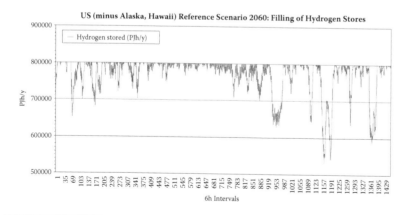

Figure 9.8 Hydrogen filling (in energy units) of all hydrogen stores assumed to be incorporated in the 2060 contiguous US reference scenario (see Chapter 4). The total capacity corresponds to about 10 days of demand from the US transportation sector, and the stores remain over 60% filled at all times during the year, being replenished mainly by surplus electricity from photovoltaic and wind power, which is rather well distributed over the year.

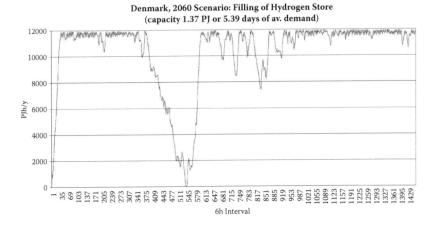

Figure 9.9 Hydrogen filling (in energy units) of the Danish hydrogen stores assumed to be incorporated into the 2060 scenario for northern Europe (Sørensen 2008b). The total capacity corresponds to about five days of demand from the Danish transportation sector, which leads to a hydrogen storage level briefly reaching zero during an exceptionally long spring lull in the main Danish power source (wind) for the particular year used in the simulation. The store is assumed empty at the beginning of the year, but is seen to be quickly filled to maximum capacity. Danish wind resources are generally largest in winter, so in practice, there would always be energy in the stores at the beginning of a year.

most wind energy during winter. The largest drawing on the hydrogen stores is in the fall, at a meteorological incidence of low wind and overcast skies over several areas of the United States.

For the northern European 2060 scenario, the hydrogen storage requirement is highest in Denmark, where a minimum of 5.4 days of average demand by the transportation sector needs to be available (Sørensen 2008b). As Figure 9.9 shows, this is the absolute minimum capacity for coping with a particularly extended spring period of low wind during the year from which the wind data have been extracted. For supply security, a more appropriate storage size would be around 10 days, as in the US case. The reason that hydrogen storage is more important for Denmark than for the other northern European countries is that the scenario has Denmark deriving its electric power almost entirely from wind and is therefore dependent on the variability and intermittence of this particular source flow. Germany has substantial photovoltaic generation, and hydro is important in all Nordic countries except Denmark. The fact that the storage demand in Denmark is still rather small is, of course, due to the persistence of winds and the substantial coverage of intermittency problems by power transmissions between the countries. The five demand-days of

Danish hydrogen storage are easily covered by the already-existing two aquifer and salt dome caverns (currently storing natural gas at a higher pressure), and additional suitable sites have been identified, ensuring that any further needed hydrogen storage capacity could be built at a modest cost.

Several technologies have been considered for short-term storage of high-quality energy forms, such as flywheels (installed underground for safety) and supercapacitors, typically aiming at removal of spikes in power delivery caused, for example, by switching between production units of power supply. Although such power quality issues may be important for certain applications, an increasing portion of the electrically powered equipment we currently employ is considerable more resilient to deficiencies in power quality (voltage excursions, cycle frequency deviations, current spikes) than the equipment used a few decades ago. Furthermore, the devices that have a dependence on power quality will now often be equipped with a battery or capacitor unit taking over in case of power interruption and generally smoothing any short-term variations in power definitions. These issues are further reduced in importance for emerging electricity systems based on renewable energy sources such as direct solar radiation or wind, because generation by a large number of dispersed units will not give rise to the power quality problems caused by the falling out of single, large generation unit.

A particular role is played by battery storage, which is in everyday use for small devices such as smart phones and portable computers, for medium-size devices such as lawn mowers or chain saws, and increasingly for larger types of equipment such as electric cars and other vehicles, with prospects for eventually making it to the electric utility arsenal for coping with intermittency. Because of the possibility of using controlled time displacement of battery charging, they are prime candidates for the demand management of intermittency, a topic that is covered in Chapter 11. However, batteries may also serve a direct purpose as storage facilities offering both short- and medium-term storage times, as they indeed did already in the early twentieth century (Figure 9.10). Because of the high current price of large-scale battery storage compared to storage methods such as hydro reservoir or underground cavern storage, batteries are not yet being used for bulk energy storage. Batteries are currently used mainly for small

Figure 9.10 Lead-acid battery store used in the early twentieth century to handle intermittency for a Danish wind-turbine installation (Hansen 1985; Sørensen 2012b; used with permission from the la Cour Museum).

devices or in special circumstances, such as in remote mobile applications (e.g., exploration or military action in areas without grid power). Several such autonomous battery applications, i.e., those belonging in the small-storage-device survey presented in Chapter 10, may receive competition from various fuel-cell technologies in the future (Sørensen 2012a).

References

Bauer, D., R. Marx, J. Nußbicker-Lux, F. Ochs, W. Heidemann, and H. Müller-Steinhagen. 2010. German central solar heating plants with seasonal heat storage. *Solar Energy* 84:612–23.

Brüel, P., H. Schiøler, and J. Jensen. 1976. Project report on solar domestic heating. (In Danish, but summary is included in Sørensen [1979, chap. 5; 2010, chap. 5].)

Chen, H., T. Cong, W. Yang, C. Tan, Y. Li, and Y. Ding. 2009. Progress in electrical energy storage systems: A critical review. *Prog. Natural Science* 19:291–312.

Díaz-González, F., A. Sumper, O. Gomis-Bellmunt, and R. Villafáfila-Robles. 2012. A review of energy storage technologies for wind power applications. *Renewable and Sustainable Energy Reviews* 16:2154–71.

Elliston, B., M. Diesendorf, and I. MacGill. 2012. Simulations of scenarios with 100% renewable electricity in the Australian national electricity market. *Energy Policy* 45:606–13.

Fujino, J. (ed.). 2008. *Low-carbon society scenarios towards 2050*. Global Environmental Research Fund and Japan-UK Joint Research Project, Japanese National Institute for Environmental Studies. http://2050.nies.go.jp.

Glasnovic, Z., and J. Margeta. 2011. Vision of total renewable electricity scenario. *Renewable and Sustainable Energy Reviews* 15:1873–84.

Haillot, D., F. Nepveu, V. Goetz, X. Py, and M. Banabdelkarim. 2012. High performance storage composite for the enhancement of solar domestic hot water system. Part II: Numerical system analysis. *Solar Energy* 86:64–77.

Hansen, H. 1985. *Poul la Cour—grundtvigianer, opfinder og folkeoplyser*. Askov, Denmark: Askov Højskoles Forlag.

Inage, S. I. 2009. *Prospects for large-scale energy storage in decarbonised power grids*. International Energy Agency working paper, Paris.

Japanese Wind Power Association. 2004. Countermeasures of electricity frequency change (in Japanese). http://www.meti.go.jp/committee/downloadfiles/g40422a50j.pdf.

Li, C., and R. Wang. 2012. Building integrated energy storage opportunities in China. *Renewable and Sustainable Energy Reviews* 16:6191–6211.

Marx, R., J. Nußbicker-Lux, D. Bauer, W. Heidemann, and H. Drück. 2011. Saisonale Wärmespeicher—Bauarten, Betriebsweise und Anwendungen. *Chemie Ingenieur Teknik* 11:1994–2001.

Pinel, P., C. Cruitckshank, I. Beausoleil-Morrison, and A. Wills. 2011. A review of available methods for seasonal storage of solar thermal energy in residential applications. *Renewable and Sustainable Energy Reviews* 15:3341–59.

Sharma, A., V. Tyagi, C. Chen, and D. Buddhi. 2009. Review on thermal energy storage with phase change materials and applications. *Renewable and Sustainable Energy Reviews* 13:318–45.

Sørensen, B. 1979. *Renewable energy*. 1st ed. London: Academic Press.

Sørensen, B. 2007. *Renewable energy conversion, transmission and storage*. Burlington, MA, and London: Academic Press/Elsevier.

Sørensen, B. 2008a. A sustainable energy future: Construction of demand and renewable energy supply scenarios. *Int. J. Energy Research* 32:436–70.

Sørensen, B. 2008b. A renewable energy and hydrogen scenario for northern Europe. *Int. J. Energy Research* 32:471–500.

Sørensen, B. 2010. *Renewable energy*. 4th ed. Burlington, MA: Academic Press/ Elsevier.

Sørensen, B. 2012a. *Hydrogen and fuel cells*. 2nd ed. Burlington, MA: Academic Press/Elsevier.

Sørensen, B. 2012b. *A history of energy*. Oxford, UK: Earthscan/Routledge.

Sørensen, B., and P. Meibom. 2000. A global renewable energy scenario. *Int. J. Global Energy Issues* 13 (1–3): 196–276. doi: 10.1504/IJGEI.2000.000869.

Sweet, M., and J. McLeskey. 2012. Numerical simulation of underground Seasonal Solar Thermal Energy Storage (SSTES) for a single family dwelling using TRNSYS. *Solar Energy* 86:289–300.

Tatsidjodoung, P., N. Pierrès, and L. Luo. 2013. A review of potential materials for thermal energy storage in building applications. *Renewable and Sustainable Energy Reviews* 18:327–49.

Tsuchiya, H. 2012. Electricity supply largely from solar and wind resources in Japan. *Renewable Energy* 48:318–25.

Yumrutaş, R., and M. Ünsal. 2012. Energy analysis and modelling of a solar assisted house heating system with a heat pump and an under-ground energy storage tank. *Solar Energy* 86:983–93.

Zhang, Q., K. Ishihara, T. Tezuka, and B. McLellan. 2012a. Scenario analysis on future electricity supply and demand in Japan. *Energy* 38:376–85.

Zhang, Q., T. Tezuka, K. Ishihara, and B. McLellan. 2012b. Integration of PV power into future low-carbon smart electricity systems with EV and HP in Kansai Area, Japan. *Renewable Energy* 44:99–108.

10
DECENTRALIZED STORAGE

Many of the techniques used for larger-scale energy storage can, in a suitably modified form, also be used in small units and for dispersed application. For example, metal or composite material containers can be used to store liquid and gaseous fuels (rather than the excavated underground stores described in Chapter 9). Flywheels have, over the past century, had small-scale applications for smoothing the operations of motors or engines at rotational speeds lower than that of the underground devices currently considered with the use of advanced materials (Koohi-Kamali et al. 2013). Heat stores based on heat capacity or phase-change energy are used in relatively small-scale contexts, and batteries are used on any scale from nail-head size to utility storage banks. If hydrogen makes it into the future transportation sector, then safe hydrogen containers for automotive applications may be furnished by high-strength composite materials, probably a more likely solution than liquefied hydrogen stores (which lose energy when stored for several days). Figures 10.1 and 10.2 show examples of hydrogen-storage placement in recent prototype hydrogen vehicles. For consumer cars, the new equipment will be encapsulated and hidden under the floor (Figure 10.2) or elsewhere, and the car will be difficult to distinguish from a gasoline or diesel cousin.

One vision of energy supply in future societies would involve dispersed generation from wind-turbine parks and building-integrated photovoltaic panels. Today, most buildings have facilities installed that will provide the necessary space conditioning by employing a heating and cooling system. Photovoltaics is one example of a technology that promises each building to also be able to furnish part or all of its electricity needs, the latter requiring building-adapted storage to be part of the concept. Generally, many citizens consider it a positive thing to be "in charge" of one's own energy supply, and the addition of electricity production to the arsenal within the dwelling is already

Figure 10.1   Hydrogen store (integrated into backseat) and fuel cell plus control equipment placed in the back of a prototype car made for the Danish Hydrogen Demonstration Program 2000 by FIAT/IRD (photo by author).

attracting interest. For example, natural gas burners are being replaced by power- and heat-delivering gas furnaces, available today in handy units that take up no more space than pure gas heaters. The flexible supply of natural gas makes it possible to achieve this without having to install active storage of high-quality energy inside each building.

The next step toward family autonomy is to be capable of providing the family vehicle's energy needs, either by plug-in electricity for electric vehicles or by an outlet from a hydrogen pipeline or hydrogen store to a fuel-cell vehicle. The latter case invites new thinking of combining hydrogen functions, thereby reducing the number of devices required for using and storing hydrogen in a decentralized way. The panacea here is that all fuel cells in principle are reversible, meaning that they can convert hydrogen into electricity or electricity

Figure 10.2   Placement of hydrogen store and fuel cell under the floor of the 2001 Necar-5 prototype car (cf. Figure 4.8), built by Daimler AG (photo by author).

into hydrogen (plus waste heat in all the processes). In past industrial applications, the optimization of fuel cells for hydrogen production (using alkaline fuel cells called *electrolyzers*) or for power production (currently mostly focusing on solid oxide fuel cells [SOFC] or proton exchange membrane fuel cells [PEMFC]) has often been achieved at the expense of the efficiency of the reverse operation. If a cell can be constructed that is reasonably efficient for both direct and reversed operation, new possibilities appear. Most current fuel cells have efficiencies around 50% for operation in both directions, but pure electrolyzers have reached nearly 90% efficiency, and the hope for future power-generating ones is 65%–70%, which is currently in sight for stationary SOFCs but not quite for the PEMFCs aimed at the transportation sector (Sørensen 2012).

Assuming a situation where a building has power-line connections with two-way metering (a feature seen increasingly and hopefully becoming standard in the future, cf. Chatzivasileiadi, Ampatzi, and Knight [2013]), but no pipeline connections for hydrogen, then the setup may be as depicted in Figure 10.3. Such a system has been simulated for Denmark in a decentralized 2050 scenario with high dependence on hydrogen and without use of transmission lines to

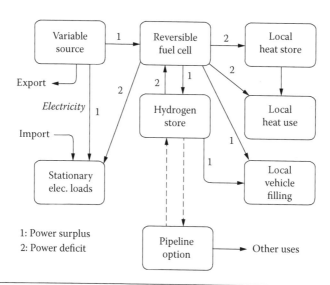

Figure 10.3 Layout of a decentralized building-integrated energy system based on intermittent primary resources and a demand of heat, electric power, and hydrogen for vehicles used by the inhabitants. Flows are indicated in the case of (1) surplus power supply and (2) supply below demand (Sørensen 2012; used with permission).

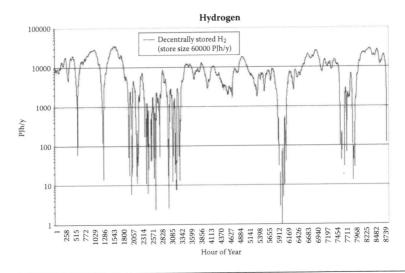

Figure 10.4 Variations in the filling of 0.3-m³ hydrogen stores (in cases where metal hydride stores are employed) attached to 2 million buildings in Denmark using fuel cells according to a 2050 scenario of decentralized energy supply and conversion (Sørensen 2012; used with permission).

neighboring countries. Hydrogen is seen as covering 72% of the mobility energy demand (of the remainder, 20% is covered by liquid fuels and 8% by electricity in trains and electric cars), and this hydrogen is derived entirely from decentralized conversion in reversible fuel cells using power produced by wind turbines (onshore and offshore) plus a little from rooftop PV panels (Sørensen et al. 2001). A centralized scenario with additional hydrogen produced from liquid fuels would, of course, show a smaller demand for hydrogen storage. Figure 10.4 shows the variations in filling of the average hydrogen store of a family building. This may be a basement or underground store based on metal or complex hydrides, with part of this storage capacity accounting for vehicles that need to be refilled with hydrogen when parked at the house. Of course, the precise amount of hydrogen stored within a vehicle will depend on its usage. It should be kept in mind that road vehicles remain parked during most of their life, either at a residence or at a working place (see the power-duration curve in Figure 10.5).

The average floor area for a 2050 Danish detached family house is assumed to be 200 m² and housing two adults and two children. Not counting vehicle storage due to the uncertainty mentioned in the previous paragraph, the single-building storage capacity is assumed

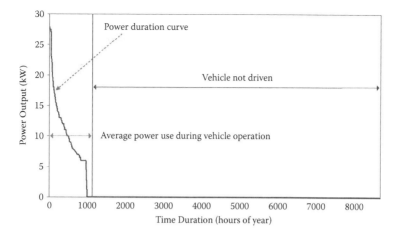

Figure 10.5 Typical annual power-duration curve for a passenger car, showing that maximum power (here 40 kW, optimal for an electric or fuel-cell vehicle) is required only rarely, that idling accounts for a couple of hundred hours a year (part of curve at 5 kW or lower), and that the car is not being driven for more than 86% of the year (Sørensen 2012; used with permission).

to be the equivalent of a 0.3-m³ metal hydride store on average (or 60 EJh for the entire country). The few hours in a year where stores become empty can be avoided by increasing the number of wind turbines, but this is hardly economic because the same effect may be accomplished quite easily by trade via the international transmission line connections.

The hydrogen storage capacity assumed in the decentralized scenario of Figure 10.4 is equivalent to 54 days of primary hydrogen demand (i.e., before conversion losses in fuel cells). This is 10 times more than in the centralized scenario depicted in Figure 9.9, which makes full use of Denmark's international power grid connections. (However, the durations are not fully comparable, because the hydrogen demand is lower in the 2060 northern European scenario, due to the larger part of the transportation requirements being satisfied by other means.) The estimated 54 days of decentralized hydrogen storage is similar to what may be accomplished in a more centralized scenario by making use of the underground caverns already available in Denmark, but the cost of establishing 2 million detached hydride stores below buildings would be a much more expensive option than the obvious alternative of using and possibly reinforcing the transmission connections to Norway and Sweden, as in the northern European scenario considered in Chapter 9.

An increasing role for decentralized energy storage is being played by battery technologies, along with the transition from the standard lead-acid batteries to various advanced concepts, notably those involving solid electrolyte lithium-ion technologies. Examples of recent progress include nanoporous electrodes (Hong et al. 2012), polymer electrode protection (Choi et al. 2013), all-solid-state lithium batteries (Hassoun et al. 2012; Unemoto et al. 2013), silver nanoparticle electrodes (Krajewski et al. 2014; Lin et al. 2014), or carbon nanotube anodes (Zhang et al. 2014). The energy management options offered by this development are further explored in Part III of this book (Chapters 11 and 12).

A large segment of electricity application is for portable computers and related equipment. These are often traveling during the daytime and docked at night, offering the possibility of recharging at a time suitable as regards diurnal demand variations, but not necessarily compatible with the variations in solar radiation. Other portable electronics such as smartphones in most cases do not need recharging every day, but again are most conveniently charged at night. The same is the case for electric bicycles (about 6-hour charging time for a 10 Ah battery), also a device with a rapidly growing market penetration. They are used for leisure, for shopping, and for commuting, all of which are chiefly daytime activities, although recharging when the bicycle is parked at a workplace during work hours is also a possibility (a battery example is shown in Figure 10.6). Garden utensils are used much more infrequently, and an off-peak, intermittent-resource adapted choice of charging periods is quite possible.

Coming to electric motorcars, they are similar to electric bicycles in that charging while parked near the workplace is feasible, as is nighttime recharging. Street-side recharging stands, including special DC chargers capable of charging an electric car with a driving range of around 150 km to 80% battery capacity in 30 minutes (as compared with 13 or 8 hours for full charging from an ordinary household outlet or from a 3.6 kW AC street unit; VW [2014]) are becoming common in many cities, with convenient electronic payment for the electricity drawn (see Figure 4.9).

For all appliances and vehicles used in a dispersed fashion, there is also a need to be able to offer recharging during periods of insufficient power production, such as those inherent in systems with intermittent

Figure 10.6 A 10-Ah electric bicycle battery manufactured in China, for operation at 36 V, giving a storage capacity of 360 Wh and a range of some 40–50 km, depending on wind and terrain (photo by author).

renewable resources. In the case of wind, lull periods are rarely longer than a week, while photovoltaic power may have some further adverse average seasonal variation (e.g., adverse as regards lighting, not cooling). If the system uses a combination of solar and wind resources (plus eventually biofuels), a few weeks worth of building-associated storage would normally suffice, or even less if suitable power-line transmission to and from adjacent areas is in place. Figure 10.7 gives an example of the hourly variation in power demand and supply from wind

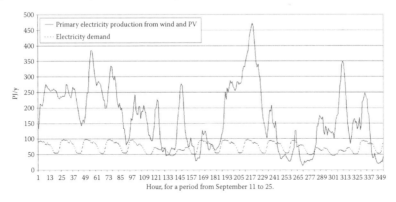

Figure 10.7 Variations in dedicated electricity demand and renewable-energy power production (also serving needs for other types of demand) for two autumn weeks in the 2050 Danish decentralized scenario with massive use of reversible fuel cells but without use of international grid connections.

for the Danish 2050 scenario considered earlier in this discussion. It indicates that deficits in electricity for dedicated applications (covering direct electricity demands as opposed to hydrogen production and heat pump operation) are rare and that stores will mainly take care of the variability in connection with other uses, provided that the usual strategy is to first serve dedicated electricity users. Other uses of electricity are primarily for hydrogen production, and the employment of hydrogen stores in the scenarios is therefore natural. The waste heat associated with all the conversions assumed to take place in individual buildings should suffice to cover hot-water and space-heating requirements, provided that the energy standards of the buildings are in the upper end, which, as discussed previously, they are indeed assumed to be in this scenario.

Technical details of many of the devices aimed at serving as energy stores may be found in review articles such as the ones quoted in this chapter and in textbooks (e.g., see Sørensen [2010] as well as the references contained therein).

References

Chatzivasileiadi, A., E. Ampatzi, and I. Knight. 2013. Characteristics of electrical energy storage technologies and their applications in buildings. *Renewable and Sustainable Energy Revs.* 25:814–30.

Choi, S., I. Kang, Y. K. Sun, J. H. Song, S. M. Chung, and D. W. Kim. 2013. Cycling characteristics of lithium metal batteries assembled with a surface modified lithium electrode. *J. Power Sources* 244:363–68.

Hassoun, J., M. Pfanzelt, P. Kubiak, M. Wohlfahrt-Mehrens, and B. Scrosati. 2012. An advanced configuration TiO_2/$LiFePO_4$ polymer lithium ion battery. *J. Power Sources* 217:459–63.

Hong, Z., M. Wei, T. Lan, and G. Cao. 2012. Self-assembled nanoporous rutile TiO_2 mesocrystals with tunable morphologies for high-rate lithium-ion batteries. *Nano Energy* 1:466–71.

Koohi-Kamali, S., V. Tyagi, N. Rahim, N. Panwar, and H. Mokhlis. (2013). Emergence of energy storage technologies as the solution for reliable operation of smart power systems: A review. *Renewable and Sustainable Energy Revs.* 25:135–65.

Krajewski, M., M. Michalska, B. Hamankiewicz, D. Ziolkowska, K. Korona, J. Jasinski, M. Kaminska, L. Lipinska, and A. Czerwinski. 2014. $Li_4Ti_5O_{12}$ modified with Ag nanoparticles as an advanced anode material in lithium-ion batteries. *J. Power Sources* 245:764–71.

Lin, C., B. Ding, Y. Xin, F. Cheng, M. Lai, L. Lu, and H. Zhou. 2014. Advanced electrochemical performance of $Li_4Ti_5O_{12}$-based materials for lithium-ion battery: Synergistic effect of doping and compositing. *J. Power Sources* 248:1034–41.

Sørensen, B. 2010. *Renewable energy*. 4th ed. Burlington, MA: Elsevier.

Sørensen, B. 2012. *Hydrogen and fuel cells*. 2nd ed. Burlington, MA: Elsevier.

Sørensen, B., et al. 2001. *Scenarier for samlet udnyttelse af brint som energibærer i Danmarks Fremtidige energisystem*. Danish Energy Agency project report (in Danish) available as *IMFUFA Text* No. 390. http://rudar.ruc.dk/handle/1800/3500, see file IMFUFA_390.pdf. (A summary is presented in Sørensen et al. 2004. *J. Hydrogen Energy* 29:23–32 along with further excerpts in Sørensen [2012].)

Unemoto, A., T. Matsuo, H. Ogawa, Y. Gambe, and I Honma. 2013. Development of all-solid-state lithium battery using quasi-solidified tetraglyme-lithium bis (trifluoromethanesulfonyl) amide-fumed silica nanocomposites as electrolytes. *J. Power Sources* 244:354–62.

VW. 2014. Presentation of e-up! *View—Volkswagen Magazine*, DK March issue, Scandinavian Motor Co., Brøndby.

Zhang, F., R. Zhang, J. Feng, and Y. Qian. 2014. $CdCO_3$/carbon nanotube nanocomposites as anode materials for advanced lithium-ion batteries. *Materials Lett.* 114:115–18.

PART III
MANAGING LOAD MATCHING

For both conventional and renewable energy systems, a central management problem is to accommodate peak demands and fading demands in an energy production system basically geared to fairly constant production, or to accommodate variable production that often is difficult to control. The word *often* is used rather than *always* because there may be other possibilities, such as for a wind turbine with variable geometry, where regulation of rotor blade angles can increase production at low wind speeds or reduce it if wind speeds are too high.

On the user side, there generally will be a natural distinction between loads that need to be satisfied at a given time and those that may be deferred if it happens to be inconvenient or impossible to cover them instantaneously. For the latter, there is usually a further categorization according to the maximum time delay that is acceptable in getting the task performed. Chapter 11 deals with these management options, both for centralized and decentralized systems such as the autonomous building systems discussed in previous chapters. In Chapter 12, the specific possibility for electric power delivery systems to perform load management by using the distribution grid to transmit messages to local systems will be explored. Needless to say, the optimum demand management strategy may not be the same for all energy systems, i.e., a strategy for a system where all components have reached a high technical efficiency would differ from that for the currently prevailing system of often deplorably low energy efficiency.

11

LOAD MANAGEMENT

A very important methodology related to avoiding peaks and other energy supply difficulties on a timescale of hours is to manage work hours. In the past, industry, office, and retail shopping had hours that were rigorously set at the same fixed start and end times for all workers, employees, and shoppers. Today, many employers offer flex-time, allowing time-shifts of working hours by up to a few hours. This has had substantial impact on the peak loads (of people and of energy) characterizing commutation to and from work by car, bus, or other means of transportation. In heavy industry with work in two or three shifts a day, this has rarely been seen as possible (although one may ask why), but globally, the fraction of all energy consumed by heavy industry has diminished in recent decades, not by becoming less important than before, but by having had less growth than the "soft" energy-use sectors. The time displacements occurring at present not only affect transportation, but also electricity and other energy demands by smoothing previous peaks seen when people were pre-paring to get to work (e.g., cooking breakfast, washing, showering) and shortly after the end of work hours (e.g., cooking dinner). The flextime schemes have smoothed energy consumption in offices to some extent, and particularly the extended-time options introduced in many countries for shopping have smoothed peaks in energy use of various kinds for this particular type of activity. Also affected in a positive way is the time variation of demands for household energy, such as heating or cooling, when people will no longer all have to be at work during the same hours.

An opposing argument is that with flexible work arrangements, homes are no longer empty during work hours (allowing the tempera-ture to be lowered), but the situation in the past may also not have fea-tured daytime unoccupied houses, due to the presence of housewives and children. Mixed advantages and, energywise, disadvantages are induced by the current tendency in many countries to increase the commuting distance, because more wealth allows families to move to more attractive dwelling locations, even if the distance from work is larger. The same is caused by the requirements of more specific skills for many complex tasks carried out in present societies: Unskilled labor may be found in the neighborhood of the workplace, but people with specialized or high education often have to be sought at considerable distance from the workplace in question.

These examples describe measures of a general nature, although they do in some cases involve decision making by individual persons. Other energy-management options try to shift loads by imposing a time-dependent price differentiation aimed at filling in consumption gaps by shifting peak-load activities to the gaps. One can argue that the price differentials should reflect the true differences in peak and off-peak energy cost, but in many cases they do not do this precisely, either because the factors influencing the price structure are too complex or because specific regulation aims to encourage higher-than-economically-warranted price signals. Such policies may be justified by referring to studies showing that price signals have to be exaggerated to achieve the right effect, because some actors will always fail to respond, at least to weak price signals, and thus would be violating theoretical assumptions of economically "rational behavior." Another avenue tested is to differentiate the price signals depending on the energy-efficient behavior of each particular customer in a preceding period (Lee and Yik 2002).

Figure 11.1 shows 24 hours of household electricity use in a monitored family dwelling in Sicily. Dishwasher and washing machine energy consumption appear at specific times, and the remaining electricity curve shows periods with energy use for activities such as cooking, light, and entertainment. The figure also indicates electric power yield from a rooftop solar PV system. Considerable load leveling could be achieved by having the dishwasher not start simultaneously with the evening television and computer game peaks, but on the other hand, the washing machine is already started at a favorable time with solar power production and few other loads. The energy system in the house depends on the metering and control equipment indicated in Figure 11.1 for accomplishing the monitoring and

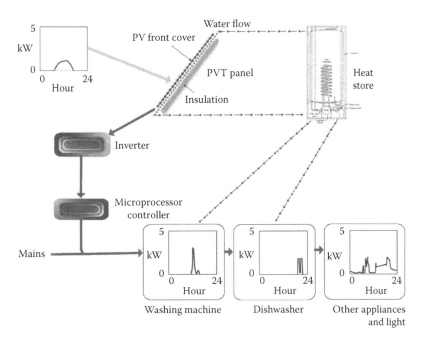

Figure 11.1 Layout of a residential power system with indication of loads and solar energy generation on a particular day, based on a setup in Sicily aimed at testing smart-grid options (data derived from Miceli [2013]). The electricity demands for a washing machine and a dishwasher have been singled out to show the limited match to available solar energy. While the washing machine is operated during sunshine, dishwashing is not, despite the operation of a microprocessor for load shifting. Evidently, it can be overruled by customer preferences. In any case, the solar panel is rated at such a low level that only a small fraction of the tasks required can be covered.

possible deferment of power-consuming activities by means of a suitable computer algorithm.

Kyriakarakos et al. (2013) combine several such buildings in a detached mini-network aimed at representing a possible energy solution for remote areas, with autonomy achieved by combining renewable power, desalination, and hydrogen technologies. Because both potable water and hydrogen can be stored, suitable sizing of the wind and solar generators will ensure full coverage of demands, but the proposed system aims to lower cost by proposing a smaller system equipped with a demand-side management system that basically just disconnects loads according to a task priority list. The simulated system is servicing a small Mediterranean island inhabited by eight people with a summer peak load of 45 kWh/day plus 1.9 m³ of clean water. Daily transportation amounted to 50 km, covered by 2.4 Nm³

Table 11.1   System Components for a Small Greek Island Energy and Water System

COMPONENT	
Number of typical 150 W$_p$ PV modules	36
Rated power of the fuel cell (W)	300
Rated power of the electrolyzer unit (W)	1000
Hydrogen storage capacity (Nm^3 of H$_2$)	12
Potable water tank capacity (m^3)	32
Rated power of the desalination unit (W)	900
Energy capacity rating of each of the 2-V batteries, a total of 24 are used for a 48-V DC bus (Wh)	400

Source: Kyriakarakos et al. 2013.

of hydrogen used in two light scooters (what interesting activities that would entail such a level of transportation needs is not disclosed). With a chosen cost ceiling below the one that would have allowed 100% coverage of loads, the optimum component size is as given in Table 11.1. The load shedding via demand management brings the number of summer days without hydrogen down to a tolerable level and allows all high-priority demands to be satisfied.

Small islands have long been scenes of interesting demand management. Fair Isle off the Scottish coast (population 74 in 20 households) had in 1982 established a power-supply setup with a wind turbine and a diesel generator. The production cost of diesel power was three times higher than that of wind power, due to the cost of sailing diesel drums to the remote island, and each house on the island had been equipped with a green and a red light. When the red light was on, power came from diesel, and when the green light was on, from wind. The cost incentive and clear information display caused a maximum shift of loads to the windy periods (Sørensen 1986). Many recent smart-grid systems of system models use a variety of algorithms to control the match between production and demand.

The Greek island example presented here used supply denial for the least important demands; other systems use time deferral of demands that cause trouble for the system resources available, and a recent Spanish proposal uses running economic selection between several conventional or renewable resources to minimize consumer costs (Álvarez-Bel, Escrivá-Escrivá, and Alcázar-Ortega 2013). The challenge for this kind of economic optimization is, of course, to make use of bound production from variable renewable generators, which is often done by assigning a price of zero at the relevant periods in time to make sure that the intermittent resources are used before any production unit that can be regulated. The catch is that the real renewable energy-generating cost is not zero, and the actual cost must therefore somehow be distributed at periods of time other than the true one. A more reasonable approach may be to regard the capital cost of all existing energy systems as "sunk costs" and then go ahead with only marginal costs and use the renewable resources as first choice for base loads, whenever their operational costs are lower than that of fuel-based units. More radical restrictions of demands that cannot be delivered by sustainable energy systems have been proposed, e.g., in Brazil (Blasques and Pinho 2012). On the software side, suggestions of load predictions based on fuzzy logic or neural-network methods have been explored to also allow system adjustments in cases where some components have extended startup or level-changing times (Álvarez-Bel, Escrivá-Escrivá, and Alcázar-Ortega 2013; Matallanas et al. 2012).

The use of demand management in 100% renewable-energy scenarios is a tool in addition to trade exchange and storage, where shifting the time of demands by a few hours may remedy situations where there is insufficient production of renewable power. Figure 11.2 illustrates such a situation by showing the disposition of power for a particular day during the season with the least wind energy production in the Danish scenario outlined in the discussions of storage in Chapters 9 and 10 (see Figure 10.3). The use of this model for general time-shifting analysis is continued in Chapter 12.

An interesting Irish study shows that buildings heated with heat pumps and including a small thermal store can switch the heat-pump power off during the late afternoon peak electricity hours without any negative effects (Arteconi, Hewitt, and Polonara 2013). Regarding peaks in heat loads, proper management of heat stores can lead to substantial load adaptation to source flows. A New Zealand example of leveling the use of electricity for space heating by installing phase-change material impregnated into office walls has been presented by Qureshi, Nair, and Farid (2011). Of course, electricity should not be used for space heating in any case except with the use of heat pumps, and properly insulated buildings at locations such as Auckland should

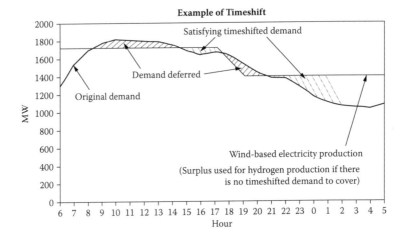

Figure 11.2 Example of using the time-shifting methodology over a 24-hour period during the Danish summer (using the simulation model to be detailed in Chapter 12). Insufficient wind power production during the day peak is dealt with by deferring demand, first to after the peak period, but as this is insufficient due to the sunset lull phenomenon, also further into the night, with maximum time shifts of 12 hours.

have very modest space-heating requirements, as shown in Figure 9.2. In rapidly developing countries such as China, traditional regulation by cutting off customers when supply is insufficient to cover demand is yielding to smarter regulation in line with the demand-side techniques being tested in the affluent parts of the world (Ming et al. 2013).

Countries relying on energy-intensive heavy industry have special challenges and opportunities in a future setting of substantial dispersed energy, particularly with regard to electricity production. Paulus and Borggrefe (2011) investigated the limited question of whether German heavy industry can provide load-leveling services to the remainder of the power market. Generally speaking, this would in many cases be feasible, because the processes of energy-intensive industries often comprise integrated energy stores, already required for the stability of the operations.

At the beginning of this chapter, the special issues of the transportation sector were presented (see also Smith [2008]). Among the many approaches to coping with this often rapidly growing sector, demand-side management occupies a prominent place, as suggested in Table 11.2. This is true not only for road transport, but also for air travel. For maritime transport (mainly of goods), the attention is mostly directed at the efficiency of energy use, because time shift of

Table 11.2 Some Measures To Reduce Transportation Demand

IMPROVE TRANSPORTATION OPTIONS	INCENTIVES TO REDUCE DRIVING	PARKING AND LAND-USE MANAGEMENT	POLICY REFORM AND PROGRAMS
Alternative work schedules	Walking and cycling encouragement	Bicycle parking	Access management
Bicycle improvements	Commuter financial incentives	Car-free districts and pedestrianized street	Campus transport
Bike/train transit integration	Congestion pricing	Clustered land use	Car-free planning
Car sharing	Distance-based pricing	Location-efficient development	Commute trip-reduction programs
Flextime	Fuel taxes	New urbanism	Comprehensive market reform
Guaranteed ride home	High-occupancy vehicle priority	Parking management	Context-sensitive design
Individual actions for efficient transport	Parking pricing	Parking solutions	Freight transport management
Park and ride	Pay-as-you-drive vehicle insurance	Parking evaluation	Institutional reforms
Pedestrian improvements	Road pricing	Shared parking	Least-cost planning
Ride sharing	Speed reductions	Smart-growth planning and policy reforms	Regulatory reform
Shuttle services	Street reclaiming	Transit-oriented development	School transport management
Small-wheeled transport	Vehicle-use restrictions		Special event management
Taxi service improvements			Transport demand management marketing
Telework			Tourist transport management
Traffic calming			Transportation management associations
Transit improvements			
Universal design			

Source: Compiled by Smith (2008) on the basis of Hensher and Button (2003). Used with permission.

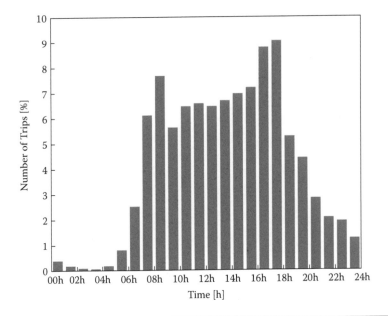

Figure 11.3 Distribution of car trips over hours of a typical day, using data from the Netherlands (Clement-Nyns, Haesen, and Driesen 2011; used with permission).

demand is unlikely to be of major importance in a business where a time delay of, say, a day, can often be tolerated. The current introduction of electric vehicles and hybrid (gasoline-electric or diesel-electric) vehicles opens a number of interesting possibilities, where the batteries in cars can be charged specifically during hours of low general demand for electricity. Typical passenger car driving patterns such as those depicted in Figure 11.3 indicate that the nighttime recharging of plug-in electric vehicles is compatible with the timing of low demand in conventional electricity systems (cf. Figure 11.2). However, another period where many plug-in electric vehicles can be partially recharged is during work hours, because their use often comprises commuting to and from work (Clement-Nyns, Haesen, and Driesen 2011).

Further opportunities offered by the expected growth in the fleets of electric and hybrid vehicles is that they can in some cases offer peak-load electricity for the power grid, or fill in when renewable power supply is low. This is true only for electric vehicles not driving at the peak-load periods, which according to Figure 11.3 will probably be only a modest fraction, but on the other hand, if delivering some power to the grid is feasible, it is not likely to negatively affect normal use of the electric vehicle, because the most favorable recharging periods lie just after the general power peak hours (Dallinger and Wietschel 2012; Tomić and Kempton 2007; Kim et al. 2012). When the mode is to have the electric vehicles act as backup for intermittent renewable sources, this may of course not be true.

In any case, the past relative neglect of demand-side options and preoccupation with expanding supply has probably come to an end, because future sustainable energy systems cannot afford to leave out half the opportunities for solving the problems. As mentioned previously, known technology can reduce demand by a factor of about five (Weizsäcker et al. 2009), and by adding smart time management of demands, the requests for a greater energy supply can be further reduced.

References

Álvarez-Bel, C., G. Escrivá-Escrivá, and M. Alcázar-Ortega. 2013. Renewable generation and demand response integration in micro-grids: Development of a new energy management and control system. *Energy Efficiency* 6:695–706.

Arteconi, A., N. Hewitt, and F. Polonara. 2013. Domestic demand-side management (DMS): Role of heat pumps and thermal energy storage (TES) systems. *Applied Thermal Engineering* 51:155–65.

Blasques, L., and J. Pinho. 2012. Metering systems and demand-side management models applied to hybrid renewable energy systems in micro-grid configuration. *Energy Policy* 45:721–29.

Clement-Nyns, K., E. Haesen, and J. Driesen. 2011. The impact of vehicle-to-grid on the distribution grid. *Electric Power Systems* 81:185–92.

Dallinger, D., and M. Wietschel. 2012. Grid integration of intermittent renewable energy sources using price-responsive plug-in electric vehicles. *Renewable and Sustainable Energy Reviews* 16:3370–82.

Hensher, D., and K. Button. 2003. *Handbook of transport and the environment.* London: Elsevier.

Kim, E., R. Tabors, R. Stoddard, and T. Allmendinger. 2012. Carbitrage: Utility integration of electric vehicles and the smart grid. *The Electricity Journal* 25 (2): 16–23.

Kyriakarakos, G., D. Piromalis, K. Arvanitis, and G. Papadakis. 2013. Intelligent demand-side energy management system for autonomous polygeneration microgrids. *Applied Energy* 103:39–51.

Lee, W., and F. Yik. 2002. Framework for formulating a performance-based incentive-rebate scale for the demand-side-energy management scheme for commercial buildings in Hong Kong. *Applied Energy* 73:139–66.

Matallanas, E., M. Castillo-Cagigal, A. Gutiérrez, F. Monasterio-Huelin, E. Caama ño-Martín, D. Masa, and J. Jiménez-Leube. 2012. Neural network controller for active demand-side management with PV energy in the residential sector. *Applied Energy* 91:90–97.

Miceli, R. 2013. Energy management and smart grids. *Energies* 6:2262–90.

Ming, Z., X. Song, M. Mingjuan, L. Lingyun, C. Min, and W. Yuejin. 2013. Historical review of demand-side management in China: Management content, operation mode, results assessment and relative incentives. *Renewable and Sustainable Energy Reviews* 25:470–82.

Paulus, M., and F. Borggrefe. 2011. The potential of demand-side management in energy-intensive industries for electricity markets in Germany. *Applied Energy* 88:432–41.

Qureshi, W., N. K. Nair, and M. Farid. 2011. Impact of energy storage in buildings on electricity demand-side management. *Energy Conversion and Management* 52:2110–20.

Smith, R. 2008. Enabling technologies for demand management: Transport. *Energy Policy* 36:4444–48.

Sørensen, B. 1986. *A study of wind-diesel/gas combination systems.* Energy Authority of NSW Report EA86/17, New South Wales Government, Sydney.

Tomić, J., and W. Kempton. 2007. Using fleets of electric-drive vehicles for grid support. *J. Power Sources.* 168:459–68.

Weizsäcker, E., K. Hargroves, M. Smith, C. Desha, and P. Stasinopolous. 2009. *Factor five.* London: Earthscan Publ.

12

USING GRIDS TO TRANSMIT INFORMATION

Load management has traditionally taken place at the end users, e.g., manual deferral of loads by switching appliances on and off. In cases where the load management is effectuated by a computerized algorithm, the data handling may in principle be taking place centrally at the electric utility company or decentralized at the user. The first solution is advocated by Chiu, Stewart, and McManus (2012). However, most investigators find it essential that the utility company cannot summarily switch off loads on the user's premises without some kind of consent (on a case-by-case basis or in terms of a more general category type of permissions). This would perhaps favor the solutions where data storage takes place locally and control is exerted by a local computer with proper information from the power provider displayed to the building inhabitants and the possibility, in specific cases, to bypass even rules that have been agreed upon generally.

In any case, such schemes will require data collection on processes and habits in each household or commercial premises, and these data would have to be transmitted to the control computer, wherever it is located. Further necessary data transport involves the information from the utility company regarding its actual and forecasted capacity factor and import options, as well as forecasts of intermittent power production over a period ranging from hours to days into the future. Furthermore, the utility company may find it useful to receive information on the totality of loads deferred by the users to better predict future demand. Technically, the assembly of all the required data from both sides can be made equally well whether centralized or decentralized. One can use the Internet, telephone, or mobile networks, or one can transmit the data (at suitable frequency levels) directly through the power lines. Data safety is an issue, because some of the data in question may be personal and sensitive. For example, data on appliance

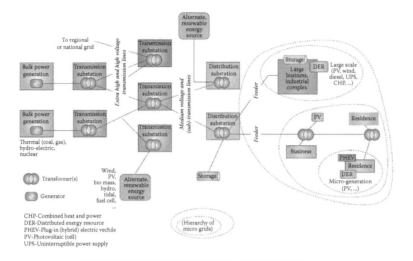

Figure 12.1   Layout of a smart grid. The dashed lines on the right-hand side of the diagram indicate various levels of local microgrids (McBride and McGee 2012; reprinted with permission of Alcatel-Lucent USA Inc.).

use and activity could be intercepted by hackers and sold to businesses that could use this information to direct unwanted advertising to the local households and firms. Questions about data security problems have been analyzed by McBride and McGee (2012), who are the source of the picture of a possible system layout shown in Figure 12.1. Increasingly detailed metering of activities on the customer side has been introduced by some utility companies in the United States, for example, with the specific purpose of optimal operation of banks of battery stores in supply systems with a variety of energy sources, including variable ones (Testa 2009).

The current trend is for recharging of small and increasingly larger batteries for battery-operated devices to take a growing share in electricity demand. This increases the options for demand management, because some amount of time displacement is possible for most of the charging tasks.

To assess the extent to which demand management can contribute to alleviating the negative effects of variable or intermittent supply, a model exercise is carried out for the energy system of Denmark, based on the scenario underlying Figures 10.3, 10.4, and 10.7, but with the following modifications: To avoid inclusion of the competing handling of supply–demand mismatch by storage of hydrogen, the use of fuel cells to generate electricity is excluded from the scenario version with demand management (but not for the base scenario used for comparison). The reverse operation of electrolyzer fuel cells to generate hydrogen fuel for the transportation sector is still included, as it justifies the large installed capacity of mainly wind power in the scenario. The original Danish model assumed a very strong curbing of electricity demand by assuming high conversion efficiency in all steps, and by invoking an attitude change to omit demands not strictly necessary. This is relaxed in the model used here, so that the Danish power demand is assumed to be rather similar (per capita) to that of the North American scenarios, with twice the reference electricity demand, as discussed in Chapter 4. The new Danish model is first run without demand management. A summary of assumptions and results for this basic scenario simulation is given in Table 12.1, and some detailed time series of the disposition of various types of energy are given in Figures 12.2–12.12.

The key features are that renewable energy sources (mainly wind and mostly offshore) are capable of supplying 90% of the electricity

Table 12.1   Summary of Basic Scenario for a 2050 Energy System Serving Denmark (PJ/y)

	LOW-T HEAT	ELECTRICITY	GASEOUS FUELS	LIQUID FUELS
Delivered energy demand	112	101	11	70
Onshore wind power potential		67		
Offshore wind power potential		213		
Photovoltaic power potential		21		
Biofuels potential				102
Solar thermal energy potential	42			
Electricity for dedicated uses		93		
Electricity for hydrogen production		108	96	
Electricity and heat from fuel cells	26	51	85	
Electricity to heat by heat pumps	72	24		
Liquid biofuels for transportation				70
Hydrogen for use in vehicles			11	
Solar thermal heat used directly	6			
Low-temperature heat from stores	10			
Discarded or lost solar heat	26			
Potential export		118		32

Note: The production potentials are values for the maximum generation equipment considered to be feasible in the scenario, and may exceed the actual production assumed.

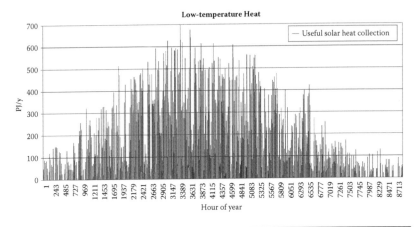

Figure 12.2 Basic 2050 scenario for Denmark. Solar thermal collection.

demand directly, plus power for heat pumps and some hydrogen production by electrolysis, as well as having an export potential similar to the dedicated domestic demand (and providing a very welcome addition to the power balance in neighboring Germany, cf. the northern European scenario in Sørensen [2008]). In the present basic scenario, the supply–demand mismatch caused by intermittence of energy production is dealt with by regenerating power from stored hydrogen. Figure 12.2 shows the photovoltaic production, which in Denmark (56° N) is poorly correlated with demand. Figures 12.3–12.5 show the heat demand and its components of coverage. Figure 12.6 shows

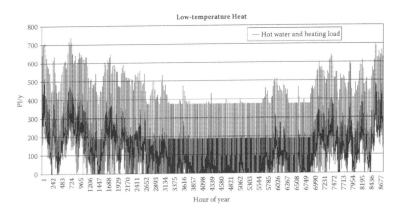

Figure 12.3 Basic 2050 scenario for Denmark. Variation of low-temperature heat loads, showing diurnal patterns of hot-water usage and seasonal variation of space heating (exhibiting the mild climate that Denmark enjoys due to the warm-water Gulf Stream near its coasts, combined with a scenario assumption of highly insulated building shells with low heat losses).

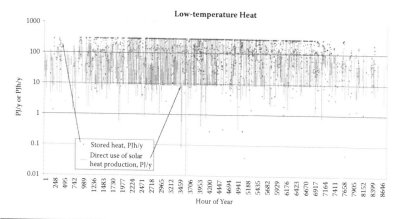

Figure 12.4 Basic 2050 scenario for Denmark. Direct coverage of heat loads by solar thermal collection and filling of solar thermal stores over the year.

electric power production and demand. Figure 12.7 shows the substantial fraction of power demands covered directly. Figure 12.8 shows the time patterns of conversion to hydrogen and heat, and Figure 12.9 the export potential, which is seen to be limited to certain hours of the year, with large gaps between them. This would be acceptable to Germany, because these imports would be a small part of the total demand. Similarly sporadic are the location of hours where the power production is insufficient and power production from hydrogen is invoked, as shown in Figure 12.10. Yet it is seen that the intermittency is fairly evenly distributed over the year, which is due to the fact that instances of poor wind conditions in no case

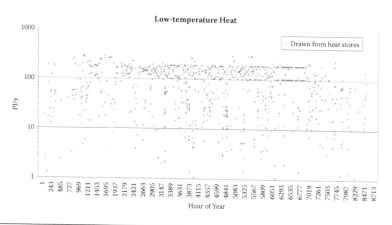

Figure 12.5 Basic 2050 scenario for Denmark. Hours where heat loads are covered from the solar thermal stores.

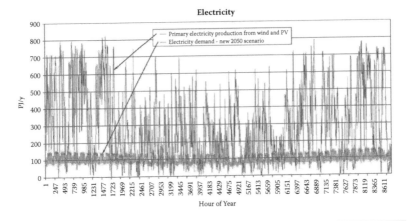

Figure 12.6 Basic 2050 scenario for Denmark. Variations in dedicated electricity demand and renewable power production over the year.

last more than two weeks during the year simulated. Figure 12.11 shows the filling pattern of the underground hydrogen stores, sized at three weeks of demand. Finally, Figure 12.12 shows the coverage of energy for the transportation sector, taken as over 80% by biofuels made from biomass residues and still being far below the potentially available bioresources from agriculture and forestry. (Danish agriculture supplies a large number of products for export, with its cereal biomass production having some 10 times the energy content of the food requirements for the Danish population.) Placing more emphasis on biofuels than on hydrogen in the transportation sector for this

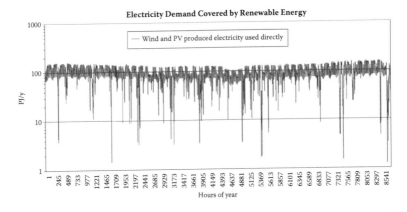

Figure 12.7 Basic 2050 scenario for Denmark. Electricity demand covered directly by renewable power production during the year.

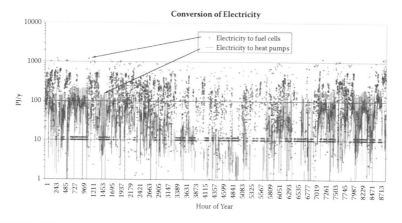

Figure 12.8 Basic 2050 scenario for Denmark. Electricity surpluses (over dedicated demands) used to produce hydrogen or applied to heat pumps for coverage of heat demands.

scenario is due to the uncertainty still surrounding the development of fuel-cell technologies (Sørensen 2012).

Omitting the option to regenerate electricity from stored hydrogen but not yet introducing demand-side management leads to increased use of liquid biofuels, as the energy system now has to supply the missing electric power at times of insufficient production by conventional power plants or combined-cycle power and heat plants, rather than from hydrogen as in Figure 10.10. The resulting balances are shown in Table 12.2, and the increased amounts of electricity available for

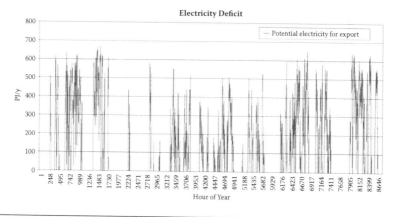

Figure 12.9 Basic 2050 scenario for Denmark. Hours of the year when potential electricity export is indicated.

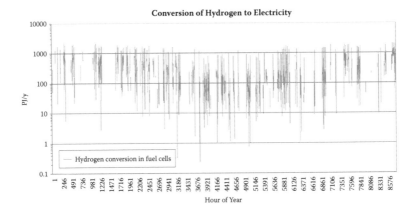

Figure 12.10 Basic 2050 scenario for Denmark. Hours in the year where hydrogen is converted back into electricity (by fuel cells) to make up for insufficient direct production.

Table 12.2 Summary of a Scenario for a 2050 Energy System Serving Denmark, Similar to That in Table 12.1, Except for Excluding the Possibility of Regenerating Electricity from Fuel Cells (PJ/y)

	LOW-T HEAT	ELECTRICITY	GASEOUS FUELS	LIQUID FUELS
Delivered energy demand	112	101	11	70
Onshore wind power potential		67		
Offshore wind power potential		213		
Photovoltaic power potential		21		
Biofuels potential				102
Solar thermal energy potential	42			
Electricity for dedicated uses		93		
Electricity for hydrogen production		13	11	
Electricity and heat from fuel cells	0	0	0	
Electricity to heat by heat pumps	72	24		
Liquid biofuels for transportation				70
Liquid biofuels for heat and power	20	8		32
Hydrogen for use in vehicles			11	
Solar thermal heat used directly	6			
Low-temperature heat from stores	14			
Discarded or lost solar heat	22			
Potential export		170		

Note: This stepwise approach makes it easier to see the particular effects of demand-side management by comparing Table 12.2 with Table 12.3.

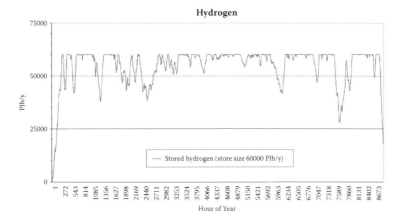

Figure 12.11 Basic 2050 scenario for Denmark. Filling of hydrogen stores over the year. The total capacity of the hydrogen stores is the equivalent of three weeks of average power demand, which is about four times higher than what was assumed in the Danish part of the northern European scenario shown in Figure 9.9. Still, it is lower than the capacity of existing Danish underground natural gas caverns, after taking into account the different energy densities and allowable pressures of the two gases.

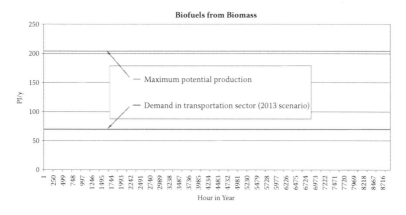

Figure 12.12 Basic 2050 scenario for Denmark. Biofuels used in the transportation sector are quite low compared to the indicated potential from Danish agricultural and forestry residues (no dedicated energy crops considered).

export in periods of surplus production, now that less hydrogen is needed, is shown in Figure 12.13.

To this model without electricity backup by forward operation of fuel cells, demand management is finally added in the following way: For the average consumer, it is considered that 10% of the power demands that cannot be covered directly by the renewable power production may be delayed 12 hours, 20% (including the previous 10%)

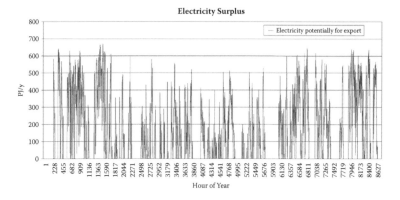

Figure 12.13   2050 scenario for Denmark, differing from the basic one by not allowing use of fuel cells to regenerate electricity from hydrogen. In consequence, more biofuels are used to cover periods of intermittent power production in conventional combined-cycle power plants (cf. Table 12.2). The hours during the year when electricity for potential export is available are shown here. The amounts are higher than in the base scenario (Figure 12.9) because domestic electricity for the hydrogen storage cycle is no longer needed, only electricity for production of hydrogen for the transportation sector.

6 hours, 30% similarly 3 hours, and 40% no more than 1 hour, while the remaining 60% cannot be delayed. The selection of which demand can tolerate a certain time shift is left to the user, but the control computer (here the simulation model) takes care of ensuring that delayed loads are covered before the expiration of the set maximum delay, doing so by imports if no other option is available before the set time limit. Table 12.3 shows the overall effect of this level of demand management. The ≈10% of electricity demands that had to be covered by biofuel conversion without demand shifts (in the model illustrated in Table 12.2) are now covered by the time-shifting strategy. This is remarkable, considering that the main power source, wind energy, exhibits a considerable short-term correlation, implying that following an hour with insufficient wind to cover demands, there is a fairly high probability that the following hours will also be characterized by low winds. However, because the direct coverage is already high, the time-shifting requirement is actually rather modest (as mentioned, on average at the 10% level), and variations in load (e.g., during the hours following peak loads) as well as in wind, even during low-wind meteorological situations, are large enough to allow demand management to solve the problem. In other geographical locations with different variable renewable-energy source flows, this may not be the case, so the conclusion from the model study presented here is that demand

Table 12.3 Summary of Basic Scenario for a 2050 Energy System Serving Denmark Shown in Table 12.1, but with Demand-Side Controls in Effect (PJ/y)

	LOW-T HEAT	ELECTRICITY	GASEOUS FUELS	LIQUID FUELS
Delivered energy demand	112	101	11	70
Onshore wind power potential		67		
Offshore wind power potential		213		
Photovoltaic power potential		21		
Biofuels potential				102
Solar thermal energy potential	42			
Electricity for instant dedicated uses		93		
Electricity for delayed uses		8		
Electricity for hydrogen production		13	11	
Electricity and heat from fuel cells	0	0	0	
Electricity to heat by heat pumps	72	24		
Liquid biofuels for transportation				70
Liquid biofuels for heat loads	25			27
Hydrogen for use in vehicles			11	
Solar thermal heat used directly	6			
Low-temperature heat from stores	8			
Discarded or lost solar heat	28			
Potential export		170		5

Note: This scenario allows electricity-requiring tasks to be deferred for up to 12 hours.

management can, at least in the present case, remedy supply–demand mismatches at the 10% level, but will likely only be able to do so partially if the intermittency mismatches are larger.

Figure 12.14 shows the disposition of surplus electricity (relative to dedicated demands) in the demand-management scenario. Relative to the scenario without demand management (Figure 12.8), the time pattern of conversion of power into hydrogen is altered, while the use of power for heat pumps is unchanged. The remaining surplus electricity for potential export is nearly unaltered from that in Figure 12.13. Figure 12.15 shows the time distribution of shifts achieved by the demand-management algorithm. Again, it is seen that the supply–demand mismatch situation under Danish conditions is fairly evenly distributed over the year, but that the interval between occurrences can be up to two weeks. There are occasions where the amount of time-shifted electricity supplied during a later hour (up to 12 hours later than the original demand) may reach an

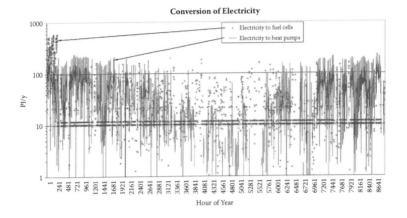

Figure 12.14 Demand management 2050 scenario for Denmark, without generation of electricity by fuel cells. Electricity surpluses (over dedicated demands) are only used to produce hydrogen (by electrolyzer fuel cells) or applied to heat pumps for coverage of heat demands. Due to the effect of demand management, the time-development pattern is different from that of the base scenario (Figure 12.8).

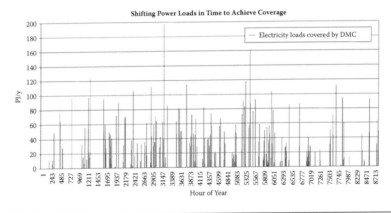

Figure 12.15 Demand management 2050 scenario for Denmark, omitting generation of electricity from fuel cells in order to show the effect of demand management more clearly. Depicted are electricity loads covered at a time shift of 1–12 hours, under control by a demand-management algorithm on a decentralized computer or a computer placed at the power utility, but in any case receiving the relevant information related to both power production and local demand.

accumulated value of nearly 200 PJ/y, which is approaching twice the maximum load in a single hour.

The Danish case study indicates that the success of demand management depends on many factors that may differ with demand and power production patterns. Making simulation studies in advance can help to identify the issues relevant for a given situation, but there is no guarantee that particular intermittency problems can be fully taken care of by shifting demand in proportions acceptable to the customers.

References

Chiu, D., C. Stewart, and B. McManus. 2012. Electric grid balancing through low-cost workload migration. *Performance Evaluation Review* 40 (3): 48–52.

McBride, A., and A. McGee. 2012. Assessing smart grid security. *Bell Labs Technical Journal* 17 (3): 87–104.

Sørensen, B. 2008. A renewable energy and hydrogen scenario for northern Europe. *Int. J. Energy Research* 32:471–500.

Sørensen, B. 2012. *Hydrogen and fuel cells*. 2nd ed. Burlington, MA: Elsevier.

Testa, B. 2009. Building the new electric grid. *Mechanical Engineering* 131 (12): 24–28.

PART IV
TRANSITIONS AND COST

Economic viability is, in a broad sense, a condition for any new system to replace an existing one, at least within the economic paradigm prevailing in most of the world today. However, this was not true for earlier, historical developments. For instance, automobiles were rapidly introduced and taking over the role of horse-driven carts around 1900, despite the much higher price tag they carried. The reasons included the fact that many ordinary people cared about the perceived difference in pollution between the smell and disposal issues of horse manure relative to the automobile exhaust, which at the time was seen as insignificant relative, for example, to that of an industry fueled by smoggy coal combustion. The question is not whether the decision to introduce fossil fuels in the transportation sector was right or wrong (in retrospect, many of us today would have preferred a development based on the electric cars, which actually were close to winning over the cars propelled by internal combustion engines), but rather to acknowledge the obvious fact that human preferences and decisions are influenced by many factors in addition to the purely economic ones. Some new technologies are inherently attractive, and customers want to be the first to possess them, even if they are more expensive than existing alternatives. The argument would often be that the high price only accompanies the introductory period, and that mass production will later lower the price—an argument that has proven right in many cases.

However, today more substantial arguments may be put forward: The first is that economic comparisons are misleading as long as "externalities" are not fully taken into account—pollution influencing human health, the environment, and the climate; the social risks connected to structural and control requirements; and the impacts from large accidents and political risks associated with, on the one hand, resource depletion and supply security and, on the other hand, the possibility of terrorist actions—all of which may be substantially different from one type of system to another. Secondly, an important problem such as energy supply has to be seen in the light of general concerns over the global societal development, where there is a growing awareness that the present economic growth paradigm is not sustainable. Nonmaterial activities can grow without limits, but material ones cannot, and the new insight dawning on many observers is that resource usage must stop long before the resources become physically exhausted. Of course, these problems are embedded in the general world political problems, where current globalization has proven to be incompatible with a power structure inviting the resolution of disagreements by use of violence and war. The division of the globe into nations may have been a good idea in the past, when fighting between tribes was a major problem that could be dealt with by furnishing a common framework of laws and conditions for cooperation. The same line of thinking today would suggest abolishing nations in favor of a global scheme based on the human rights charter already formulated by the United Nations, but with a mechanism for enforcing these and other common rules without the representatives of local communities voting according to narrow national interests, as happens now in current international organizations.

Another deviation from economic rationality must be mentioned. Currently, measures for using known technology to minimize the energy expended for a given task or benefit have a much lower status than the establishment of new energy-supply technologies (of any type). This is a plain misunderstanding of the concept of growth. It is not positive that energy use grows just because we do not implement known technologies to improve efficiency. In any respect, we would be better off by first investing in energy-efficiency measures that are cheaper than those of increasing production capacity. This would provide the same goods and services as today with four to five times less energy (Sørensen 1982, 2008a; Weizsäcker et al. 1997, 2009), making the options for a transition to a more sustainable energy supply system much wider and easier to realize.

"Economic rationality" is in fact one of the conditions that Adam Smith (1776) used for the model of a market economy, which underlies current liberal or capitalistic economic paradigms. The other two conditions for the model to be valid were that all players had approximately the same size and clout (certainly not fulfilled today, where multinational enterprises compete with small family businesses) and that all necessary information should be available to all players, because obviously without such information they could not make rational decisions (rational meaning here conformal with the selected economic paradigm). Today, businesses do everything they can to prevent others from getting at the information that would enable rational economic behavior. The result is that business decisions are rarely based on economic rationality, which may in fact not be a bad thing, as evident from the mentioned example of bold decisions that have led to entirely new products irrespective of cost. (Still, many other business decisions are based on normative positions of a rigidity even higher than that of economically rational decision making.) The bottom line is that current decision makers (in government or in business) make reference to economic theory only in the hope of avoiding debate, not because their decisions are those that would be dictated by economic rationality. This short historical excursion puts the choices of investment in new energy production equipment or efficiency measures—by private individuals, by business executives, or by parliaments and governments—into a sober perspective.

Despite the uncertainties in estimating the full cost of things, including estimation not only of the direct monetary value of materials, machinery, and labor entering, but also of the indirect costs to the extent that they lend themselves to evaluation in monetary terms (Sørensen 2011a), costs are still extremely important for people living in the present type of societies, where most necessities have a price tag and where selling personal work power is the only means of securing money for a large fraction of the population. Alternatives to this setup—in which transformation of industrial processes to use less manpower leads to shorter work hours rather than to unemployment, and with the associated changes in the distribution policies that will maintain welfare—have been proposed (e.g., see Sørensen [2012b]), but these initiatives are not close to adoption by any present society. So it may be surprising that so few costs are quoted for the technologies surveyed in this book, but the point is, as the discussion here suggests, that the true costs of both the relevant emerging technologies and the presently used technologies that have to be replaced are extremely difficult to estimate. For example, shale oil appears to be economically viable if externalities are neglected, and wind power is in many locations competitive with coal power both in terms of current market prices and certainly if externalities are included. Photovoltaics is often not economically viable, but it is making progress each year. Biofuels are perhaps competitive in a life-cycle perspective with externalities, but at least the second-generation biofuels (based on residues and not on energy crops) are not yet economically viable relative to instantaneous fossil fuel prices, if externalities are omitted and if price increases of fossil fuels over the lifetime of the user's equipment are neglected. In summary, only the present direct costs can be stated with reasonable certainty (Sørensen 2011a). Projected future costs rely on general assumptions regarding operational costs and the belief that mature technologies will have lower price tags than early versions. This may be generally true, as Figure 13.1 indicates, but exactly how much lower the price tag is going to be cannot be predicted with much accuracy.

A central issue in comparing technologies is the time distribution of costs, such as investment costs and operational costs (including fuel cost if fuels are used). Economic present-value calculations are used to make costs incurred at different times commensurable, but that depends on the interest rate used. Interest rates are parameters that can vary considerably according to the context: Are they society-wide standards used by the national banks to discount future payments? In that case, typical values over the past century are about 3% per year. Or are they personal interest rates like the ones paid on real-estate loans? In this case, they reflect our finite life expectancy and the value of possessing an asset now rather than in some distant future (where you may be dead). Interest on such personal loans are on the order of 10% or more, depending on the security the borrower is able to provide. Finally, there are intergenerational interest rates, reflecting how the present society values the future society of our children or grandchildren. An intergenerational interest rate of zero means that current and future societies are valued equally. So, for comparing two energy systems with different payment profiles, one must choose the proper interest rate, which for energy systems would typically be either the National Bank's discount rate or the intergenerational interest rate. Beware of comparisons that use a higher interest rate, as they are designed to shut out technologies with high capital cost and low operational cost, such as that of renewable energy systems. In fact, a high up-front payment may rather be seen as an advantage, as it removes much of the uncertainty of costs later in the life cycle of the technology.

Furthermore, one has to consider a very important parameter in technology comparison, that of product lifetime. Cars for personal transportation today have a lifetime in excess of 15 years, but if you are offered an electric or fuel-cell car, you should be prepared to replace the batteries or fuel cell assembly (at a high fraction of the total cost) several times during the life of the car, because these items have lifetimes of 5 years or less. Over the recent one to two decades, consumer goods manufactured in China and surroundings have penetrated the European and North American markets to such an extent that the traditional manufacturers have been forced to close their operations or to transform them into import enterprises. Consumers have been attracted to the Chinese merchandise due to a 20%–40% lower price, but they have failed to do the proper lifetime comparison with the now disappearing alternatives. In Europe, guarantee periods of 5 years have been offered by many manufacturers (and still are, e.g., by the automobile manufacturers), and consumer goods such as refrigerators, washing machines, etc., have physical lifetimes of over 15 years, like the cars. But several of the current Chinese goods have a lifetime that hardly exceeds the warranty period, typically just 1 year (several European countries have recently lowered the legal guarantee period from 2 to 1 year). In the United States and Southeast Asia, warranty periods have been as low as 3–6 months, again favoring sale of merchandise with short lifetimes, due to poor consumer awareness and uninformative advertising. The proper consumer evaluation would of course compare European and American technology with the sequence of Chinese product purchases needed to serve the consumer's needs over a similar total period, and such an evaluation would, in many cases, come to the conclusion that the "cheap" Chinese product is in reality more expensive by factors as high as 4 to 10.

The implication for assessments of new energy technologies is that one should renounce falsely precise cost figures and be satisfied with indications of life-cycle costs that are affordable in terms of the efforts made by societies, i.e., the assets required to be set aside for maintaining the energy sector as a fraction of the total level of activities. As stated, the economic estimates still do allow a number of conclusions to be drawn with little uncertainty: Invest in energy-efficiency measures until the energy system is so good that further improvements are more expensive than providing new supply. Consider not only the purely technical energy efficiency, but also the questions of whether we could arrange our settlements, activity patterns, and aspirations related to goal satisfaction in ways that use fewer resources and have a smaller impact on the environment than the ones practiced today. And finally, move sustainability up on the global agenda, so that it becomes the standard compatibility requirement for all economic and political decisions. These simple but profound changes in our behavior will require habits and mindsets to be changed, particularly in relation to economic and political paradigms. The changes will not and should not happen overnight. Rather, they should be allowed to follow a smooth path where people have time to adjust to the changes and preserve participation in the political processes of change but, of course, without blocking the necessary changes by adherence to obsolete past ways. Democracy is more than casting votes every four years on the basis of political advertising campaigns. It comprises solidarity and respect for minorities, but it also demands that citizens clinging to past unsustainable rules must give up on such habits in the absence of rational arguments against proposed changes. Clearly, these requirements suggest the need for a delicate reevaluation of the way democracy is being implemented, notably the balance between direct democracy (such as referenda) and indirect, representative democracy (political parties). Moreover, the rules of play for political campaigning and election processes ought to be aimed at broadly ensuring maximum knowledge of the implications of voting in particular ways while minimizing the leeway for nonfactual campaigns to influence the voters.

13
SYSTEMIC TRANSITIONS

Several of the simulation studies quoted in the previous chapters are examples of systemic energy transitions away from fossil and nuclear sources and toward energy supply based entirely on sustainable, renewable sources. They deal with the situations in northern Europe and North America, i.e., the regions where such transitions already have a considerable history of debate and the beginnings of serious actual implementation (e.g., photovoltaic panels in Germany and wind turbines in Denmark). Similar studies have been made for southern Europe and North Africa (Sørensen 2011b). However, the challenge of exploring sustainable futures for the growing economies in Southeast Asia has not been undertaken in depth, and a first approach is attempted here for China (in Section 13.3). Also, the Fukushima accident has revived interest for renewable-energy solutions in Japan, and Section 13.1 adds a contribution to this important case. Both examples are challenging due to the high population densities of the regions in question, and in the Chinese case, there is a unique opportunity to embark on a path leading to stable, long-term solutions without the detour of repeating the past patterns of the Western world behavior based on transitional energy solutions. South Korea is considered in Section 13.2, with an eye on the possible advantages derived from power exchange between Japan and Korea.

A central theme in treating emerging economies is population density. Historically, modernization has nearly always led to declining death rates and thus a gap between (more or less constant) birth rates and the declining death rates (Sørensen 2012b). The associated population explosion is eventually halted only when the mindsets of people have adjusted to the new reality and the realization that having many children is no longer a necessity. The latter issue is, of course, intertwined with how rapidly the new society is capable of providing care for its elderly people, so that they are not depending on

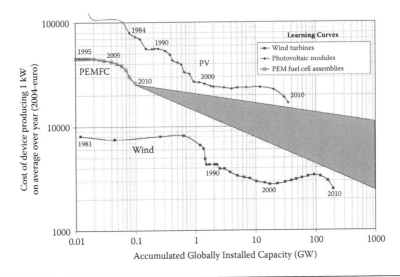

Figure 13.1   Cost of solar-cell, fuel-cell, and wind-turbine technologies as functions of accumulated volume of production. The gray area extension for fuel cells shows the behavior if learning rates turn out to be similar to either photovoltaic or wind (Sørensen 2012a; used with permission).

having some surviving children that can be persuaded to help them. In countries such as India, this type of mechanism has failed, due to a large nonurban population with little education but strong traditions for demanding care from their children. In communist China, community responsibility for social care at least for a while supported the government's one-child population-reduction policy. In Japan, overpopulation during the early twentieth century was dealt with by invading neighboring countries (e.g., Manchuria), but after the World War II defeat, a stabilization of population was promoted by education with high aspirations combined with social pressures for families to accept high expenses in bringing up children, with an ensuing reduction in birth rates. Still, the Japanese population is higher today that it was at the end of the war with its losses of life, but has again started to decline. Figures 13.2a,b show the UN projection of population size in Southeast Asia for 2050 on a logarithmic and a linear scale, respectively. This population size will be used to model the energy demands in the simulations that follow. The situation is better than in some other parts of the world, where governments deny obvious overpopulation not only to avoid providing care for the elderly, but also to breed soldiers for future wars or just to promote proliferation of specific religious institutions. Figure 13.3 shows current fertility

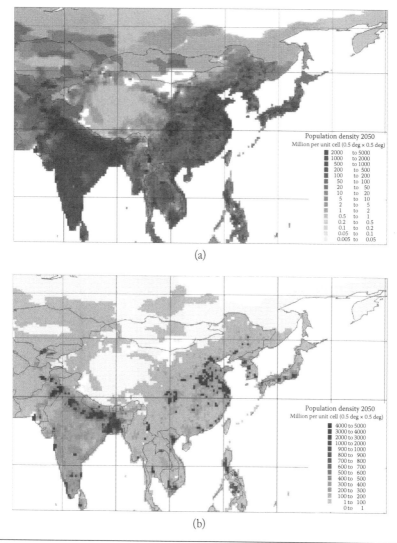

(a)

(b)

Figure 13.2 (a) United Nations projection of population density 2050 (note logarithmic scale), with modeling of migration between nonurban and urban areas added (based on Sørensen and Meibom [2000]). The figures in this chapter use the Behrmann plane projection of the Earth sphere, which is area preserving and not depending on the selection of a "center," as is the Mollweide projection that causes meridians not to be orthogonal to the equatorial parallels. The Mollweide projection is putting both "sides" of the globe on the same flattened side and is therefore used in several previous chapters of this book for global overview pictures, but for countries away from the zero-longitude meridian, the Behrmann projection is more readable. (b) Same as Figure 13.2a, but on linear scales (with a change in intervals at 1 and at 1000 million inhabitants per map cell). The map cell corresponds to about 2500 km² in central China, but varies with latitude. The lowest population areas (lightest gray) are not exactly unpopulated, as 1 million per grid cell corresponds to some 300 people per square kilometer.

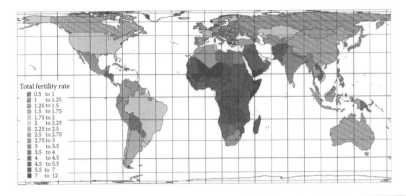

Figure 13.3 Current fertility rates for each country (based on data from United Nations Population Division 2005, latest publication UN [2013]).

rates for the countries of the world, and Figure 13.4 gives the attitudes of governments toward the situation in their own countries. Many political leaders appear satisfied with a birth rate above two or even complain that that is too low (Kazakhstan).

Another set of general conditions for energy systems is related to the height profiles of land and sea topology. Figure 13.5 shows the elevation (negative for sea floors) over standard sea level for Southeast Asia. Elevation has an influence on the availability of hydropower, biomass growth options, wind regimes, and solar radiation (through its influence on cloud cover and air turbidity) as well as technical importance, e.g., for construction of elevated reservoirs and offshore wind-turbine farms, currently restricted to foundation depths in water of about 50 meters.

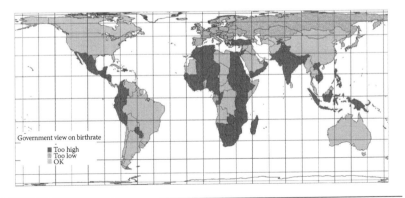

Figure 13.4 Individual government views on the fertility rates of their countries (based on data from United Nations Population Division 2005, latest publication UN [2013]).

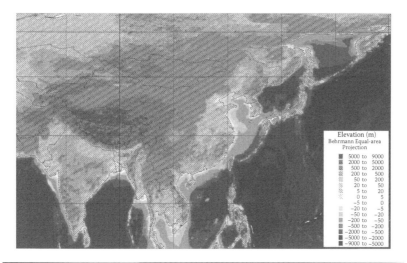

Figure 13.5 Land elevations and sea depths in Southeast Asia (m) (based on NOAA [2004]).

Figures 13.6–13.9 give the estimated annual average flux for the most important renewable energy resources of Southeast Asia—biomass, wind, hydro, and direct solar radiation—with more detail than the overview in Chapter 3. The sustainable usage for energy provisions will be discussed in the following subsections dealing with each country modeled, but a few general observations covering the entire Southeast Asian region will be made at this point. The biomass model shown in Figure 13.6 indicates that the large dry region in northwestern China is unproductive at present, but could provide considerable amounts of biomass if massive irrigation were to be introduced. This is clearly difficult with use of local groundwater, even with sparingly root-targeted irrigation techniques, because it would upset the water cycle and lower the groundwater table, as has been experienced in similar zones of Australia. However, an alternative may be solar desalination of seawater pumped in through pipeline systems (at a cost). Figure 13.9 shows that solar radiation in the area is substantial, so that the scheme appears feasible. A similar approach has already been used in Spain, where areas in the dry inland have been transformed into productive agricultural land by use of desalinated water (at first using fossil energy input but in recent years increasingly solar desalination). As in the Chinese northwest, the water balance in the Spanish highland does not permit use of groundwater. It should be noted that this kind of irrigation by imported seawater could also

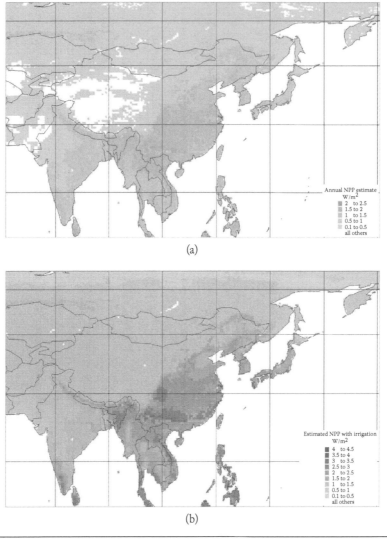

(a)

(b)

Figure 13.6   (a) Annual biomass growth, taken as net primary production in energy terms (W/m²), for the actual soil conditions prevailing and (b) for a hypothetical situation with enough irrigation to counter any water deficiency. The irrigation case cannot be considered realistic globally, but indicates what may be possible in limited areas and making use of only a modest fraction of the water cycle (Melillo et al. 1993; Sørensen 2010).

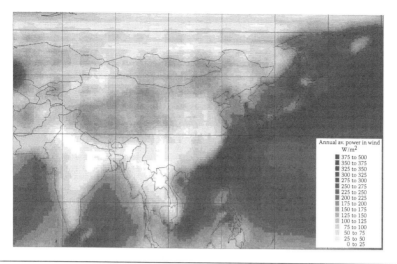

Figure 13.7 Average annual wind power that may be produced by wind turbines of currently common characteristics (W/m^2 swept for 2000, based on NCAR [2006] with use of the models of atmospheric wind conditions and of wind-turbine behavior described in Sørensen [2010]).

change the water cycle, as the water is eventually evaporated or sieves down, and in the case of new plant growth, the water cycle would be affected by changing the moisture in the lower atmosphere. With a permanent plant cover (e.g., including trees) rather than seasonally harvested crops, the entire climatic classification of the area may be changed, but fortunately in the opposite way of the earlier desertifications responsible for many current desert areas (e.g., Sahara in Africa, Rajasthan in India) through a combination of manmade processes such as the wood-removing and wood-burning practices, with subsequent evaporation and moisture changes. This is also the mechanism behind current expansion of the Sahara Desert toward the south.

The potential wind production in Figure 13.7 is estimated by use of 2000 blended land and ocean wind data in the same way as the ones used in Chapter 4 (Sørensen 2008a, 2008b). The time series of wind data are "blended" by combining results from global meteorological circulation model calculations (adjusted to measured wind data on land) with scatterometer satellite data for ocean winds. To date, this is the most accurate model available for offshore wind, because direct measurements are few. All data have been folded with the power curve of currently typical megawatt wind turbines with a hub height of around 80 meters. Excellent agreement is obtained with power production that would result from (the few) existing measured

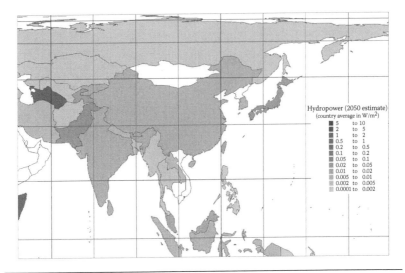

Figure 13.8 Estimate of possible hydropower exploitation averaged by country in 2050 (based on data from WEC [1995]).

wind data at this height when (a) the scatterometer data (consisting of radar reflection data received from signals sent by a special satellite) are interpreted as coming from small water droplets moved by the wind some 3 meters over the average ocean surface and (b) the wind speeds are extrapolated to 80-meter height by standard methods (cf. Chapter 3).

Looking at the resulting potential wind-power resources shown in Figure 13.7, it is seen that Japan and Korea have very substantial wind resources both on land and offshore. China, on the other hand, has decent wind resources only in a small region in the northeastern corner of the country, with some extended but more modest winds on the high plains in the northwest. However, offshore there are large wind resources off the main coastline between Hong Kong and Shanghai. It is clear even before having made any detailed calculations that a 100% renewable energy scenario for China is impossible without exploiting these offshore wind resources. In Japan, there has also recently been focus on offshore wind (Ushiyama et al. 2010), so it is important to be able to model the behavior of this energy contribution correctly. In Chapter 4, a fairly simplistic model was used for modeling offshore wind energy in North America, including only offshore supply from grid cells with both land and water content. The grid cells for North America, and the ones to be used for Southeast Asia as well, are $0.5°$

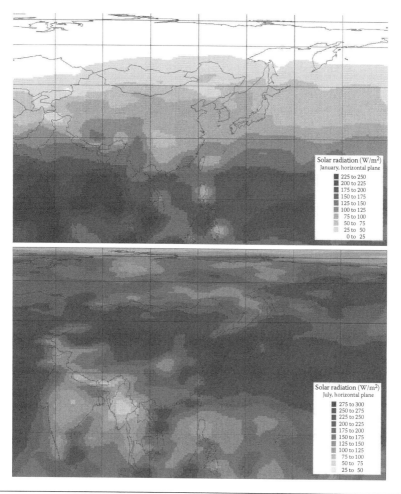

Figure 13.9 Solar radiation on a horizontal plane in January (top) or July (below). Calculated with use of 0.5° × 0.5° radiation data for 2000 provided by ECMWF (2014). (The new figures in this chapter are from Sørensen [2014a, 2014b]; used with permission.) At low latitudes, the radiation is higher in January than in July, where monsoon rain clouds obscure the sun.

× 0.5° (some 56 km × 56 km times the cosine of the latitude angle), so on average this may imply including offshore sites out to about 25 km, but in an irregular fashion not depending on actually suitable sites. This was judged sufficient for North America, where the main renewable resources are not offshore, but in Southeast Asia the situation invites a more realistic model, in particular taking into account suitability for turbine foundation at the ocean bottom. What is then done is to use the height profiles (Figure 13.5), which are available on much finer grids from the 1 arc-minute to the 5 arc-minutes used here

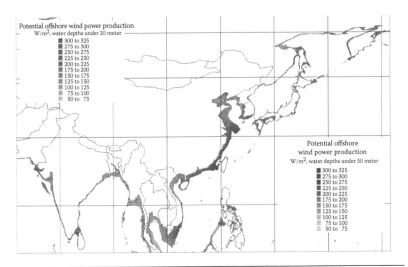

Potential offshore wind power production
W/m², water depths under 20 meter
■ 300 to 325
■ 275 to 300
■ 250 to 275
■ 225 to 250
■ 200 to 225
■ 175 to 200
■ 150 to 175
■ 125 to 150
■ 100 to 125
■ 75 to 100
■ 50 to 75

Potential offshore
wind power production
W/m², water depths under 50 meter
■ 300 to 325
■ 275 to 300
■ 250 to 275
■ 225 to 250
■ 200 to 225
■ 175 to 200
■ 150 to 175
■ 125 to 150
■ 100 to 125
■ 75 to 100
■ 50 to 75

Figure 13.10 Detail of the wind-power production map in Figure 13.7, singling out offshore resources available at water depths under 20 m (dashed) and under 50 m (solid).

(some 9.3 km), to single out the areas of the ocean where the water depth is under 20 m and under 50 m. Current offshore wind turbines, such as those in the several offshore wind parks in Denmark, have foundations stretching to depths of near 20 meters, but engineering estimates consider it technically feasible to go to depths of 50 meters, albeit of course at a higher foundation cost.

The Southeast Asian offshore wind-production-exploiting resources to 20- or 50-m depth are singled out in Figure 13.10. The procedure is as follows: The potential resources are assumed to involve an area swept by the wind turbines, which is around 0.001 (0.1%) of the ocean surface area for locations with a suitable depth. This is higher than the 0.0002 (0.02%) swept turbine area relative to land area that in Chapter 4 was assigned to wind production on land, selected with consideration of the many other uses of land areas, e.g., for habitation, agricultural and other industry, commerce, and infrastructure (roads, train corridors, airports). Offshore, other area uses are chiefly for ship routes and for fishing, of which at least fishing is not seriously harmed by the presence of wind turbines (assuming modern, environmentally acceptable fishing methods). Thus, the average maximum wind turbine density associated with the 0.1% area ratio, which works out to around one turbine per 15 square kilometers, should in most cases be acceptable. Figure 13.11 and Table 13.1 show that these assumptions

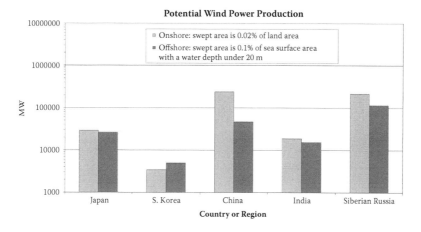

Figure 13.11   Potential onshore and offshore wind potential in selected Asian countries for the extent of employment suggested for the North American scenarios in Chapter 4. The possible need to exceed this level of exploitation, particularly for onshore resources, and the variations in production through the year, are discussed in Sections 13.1–13.3.

lead to a quite substantial offshore wind potential in Southeast Asia. Table 13.1 shows that the way of estimating wind-power potential sketched here gives considerably higher potentials than the simplistic model used in Chapter 4, except for South Korea (similar results) and Japan (lower potential, due to the rapid ocean floor drop with distance from shore, where the mixed cells in some cases extend to depths over 50 m). Table 13.1 also shows that a large part of the potential is already available at foundation depths below 20 m. In the Japanese case, fairly little is gained by going to depths of 50 m, but the situation is different in China, where the ocean areas available for wind production in some cases stretch several hundred kilometers away from the

Table 13.1   Overall Potential Wind-Power Production in Selected Asian Countries

LOCATION, DEPTH OF OCEAN USED	POTENTIAL WIND-POWER PRODUCTION (MJ/COUNTRY)				
	JAPAN	SOUTH KOREA	CHINA	INDIA	SIBERIAN RUSSIA
Offshore, mixed cells only	94,480	12,113	36,199	16,500	98,613
Offshore, to 20-m depth	26,395	4983	46,376	15,428	115,545
Offshore, to 50-m depth	35,732	11,125	94,859	25,413	173,754
Onshore, 0.02% swept	28,879	3435	240,412	18,642	214,322

Note: Values are based on the use of a rotor swept area of 0.1% of the surface area offshore and 0.02% onshore. Siberian offshore locations are mostly in Arctic waters, where turbine erection may be difficult.

shore. Details of the wind resources used in the scenarios are given in the subsequent sections on each country.

The hydro potential given in Figure 13.8 is largely from existing facilities, considering that new large-scale hydro is difficult to make environmentally acceptable. A certain expansion in small-scale hydro projects is envisaged in schemes where negative effects can be minimized.

The solar energy potential for a horizontal plane shown in Figure 13.9 is based on satellite data and calculations estimating the ground-level radiation after correction for reflection on the surface or from clouds and particulate matter in the atmosphere (ECMWF 2014). The data exists as a time series with an interval of six hours, which is likely to give more accurate results for solar utilization than the simplified model used for North America in Chapter 4. Similar to the North American calculation, transformation of the resource data to tilted surfaces was not attempted because the available model (Sørensen 2010, chap. 3) is considered too uncertain for estimating detailed geographical distribution of radiation, due to its treatment of scattered radiation in a globally averaged way. There are only a few Asian measurements of solar radiation on inclined surfaces. Those existing are for city locations and are used by architects and engineers for designing buildings, and they are underlying the "reference meteorology data" compilations used in the calculation of heating and cooling needs for individual buildings. In contrast, meteorology and climate circulation models work with horizontal-plane data alone and thus have not spurred the scientists involved to think of ways to globally measure or calculate the solar radiation on tilted planes. The optimum tilt angle for solar collectors in most of China, Korea, and Japan would be around 30° S (roughly equal to the latitude), but the radiation does not change greatly from 0° (horizontal) to 30°, so the collector performance calculation made with the horizontal data will be approximately correct. In practice, using a nonzero inclination has advantages related to accumulation of dust on the collector surface, but in any case, the precise mounting angle is determined by other factors, at least in the case of building-integrated solar panels. The 45° inclined roofs common in Europe are not prevalent in Southeast Asia, while flat roofs are more common and will allow solar panels to be mounted at any convenient tilt angle, as is also the case for centralized installations on marginal land (e.g., the large desertlike areas in north-western China). The data used for constructing Figure 13.9 have the resolution given by a grid-size of 0.5° × 0.5° (some 56 km × 56 km at low latitudes) and therefore have more details than the global data on a 2.5° × 2.5° grid used for Figure 3.2. For Japan and Korea, the scenarios to be worked out in Sections 13.1 and 13.2 use solar data on a still finer scale, given by a grid of 0.125° × 0.125°.

The future energy demands that the identified resources must be able to cover are discussed in the individual country sections, but a few general observations covering more than one country will be given here. Figure 13.12 shows the space-heat demand in Asia for 2050, using the population projections of Figure 13.2 and (as in Chapter 4) the relation

$$P = cd\Delta T$$

giving the heating power (or energy flux) P (in W/m^2) demanded at a temperature difference ΔT taken as 16°C minus the outside temperature (assuming that some 4°C heating is provided by activities in the building), and with d being the local density of population (capita per m^2) and c is a constant, which for the average building standard assumed here equals 24 W per capita per °C (consistent with 2005

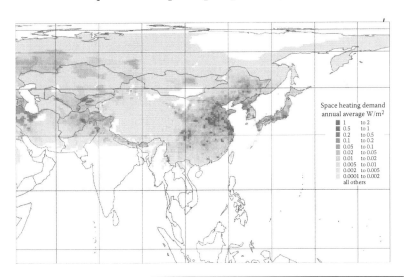

Figure 13.12 Average 2050 space-heating demand W per m^2 of land area in Southeast Asia, assuming well-insulated buildings with an average floor space of 60 m^2 per capita and the UN projection for population densities (determining the volume of buildings to be heated).

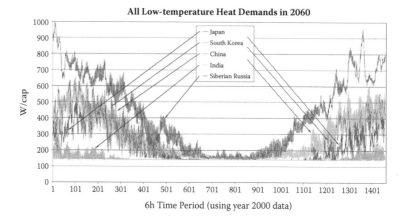

Figure 13.13 Time series of space- and water-heating demands (as Figure 13.12 using 2000 temperature data from ECMWF [2014]) in selected Asian countries, expressed as W per capita assuming an average floor area of 60 m² per capita, comprising both home, work, and other buildings.

building codes imposed in Denmark, lower than the one characterizing the existing building stock but higher than that of the most recent building code prescriptions, cf. Sørensen [2010]). Figure 13.12 thus folds the presence of low temperatures with the number of people residing at a specific location. Figure 13.13, based on the same calculation, shows the time development of per capita low-temperature heat demand over the year, now adding hot-water provision (at 150 W per capita) to the space heating depicted in Figure 13.12 and summing the data over each country. The obvious seasonality of these demands constitutes the major challenge for renewable-energy heat supply, particularly if solar thermal systems are employed.

Figure 13.14 gives the geographical distribution of space-cooling demands for the same area as the heating demands in Figure 13.12. Again, the data reflect both population densities and high temperatures. The same relation $P = cd\Delta T$ as for heat is used, but now with the temperature difference being equal to the outside temperature minus 24°C, considering that the optimum comfort temperature is about 20°C but that the first 4°C above that can be managed by using lighter clothes (as the body temperature is around 37°C, the "comfort temperature" of 20°C already assumes that clothes are worn in an amount of about two layers). Again the constant $c = 24$ W per capita per °C used reflects current ambitious building codes rather than historical building practices, which would seem an appropriate

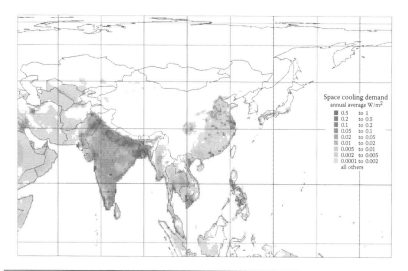

Figure 13.14 Average 2050 space-cooling demand W per m² of land area in Southeast Asia, assuming well-constructed buildings (e.g., not glass facades with excessive absorption of solar radiation) present in proportion to population density.

assumption for a scenario pertaining to a future date several decades away. Figure 13.15 gives countrywide per capita time sequences of cooling demands, being summer-concentrated except for the very large cooling requirement in India, which has a broad spring peak and a lower fall peak due to the monsoon-dominated climate. If cooling is to be accomplished by an electricity-powered compressor unit, the electricity input is some three to five times lower than the energy

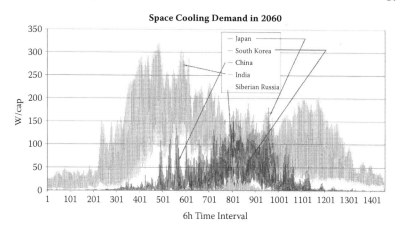

Figure 13.15 Time series of space-cooling demand (as Figure 13.14 using 2000 temperature data from ECMWF [2014]) in selected Asian countries, expressed as W per capita assuming an average floor area of 60 m² per capita for both home, work, and other buildings.

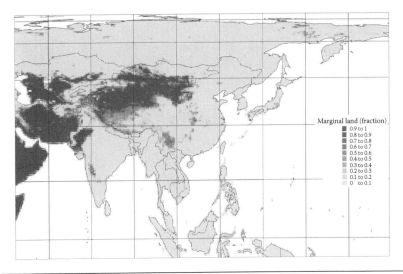

Figure 13.16 Occurrence of marginal land, expressed as the fraction of each grid cell considered marginal by international geophysical definitions (data from collection at Columbia University [2013]).

value of the heat moved from the building to its surroundings to achieve the space cooling.

Because extraction of renewable energy from marginal land areas will play an important role in the scenarios constructed (for placement of wind farms, central solar collector arrays, or growth of special desert-adapted plants), Figure 13.16 shows the occurrence of marginal land in the region under study. While the presence of marginal land is minute in Japan and Korea, it is large in China because of the high-altitude plains in the northwestern region. The cropland distribution depicted in Figure 13.17 naturally occupies areas quite different from those with marginal land.

The following scenarios use biofuels and hydrogen fuel cells in the transportation sector. As in Chapter 4, the fuel-cell vehicles could equally well be hybrid (electric–fuel cell) cars, replacing some of the vehicle fuel-cell capacity by batteries and using some of the electricity to charge batteries rather than to produce hydrogen. This will not make significant changes in the amount of electricity surpluses transferred to the transportation sector, because both battery charging and electrolyzer efficiencies are around 90%. However, the battery-discharging efficiency is higher than that of the fuel cell, so there may even be a lowering of the scenario energy input requirements.

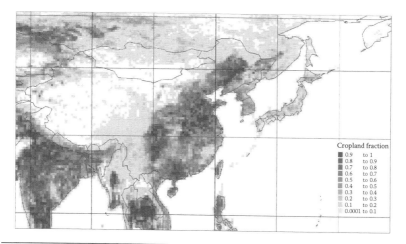

Figure 13.17 Distribution of cropland, expressed as the fraction of each grid cell used for agricultural crop production (data from collection at Columbia University [2013]).

13.1 A Japanese Renewable Energy Future

The present population density in Japan is high when seen in relation to the indigenous natural resources in the country. Japan currently provides welfare to its population by way of a large export industry, selling quality consumer goods (e.g., in the field of electronic devices) globally and thereby earning the money required for import of lacking food and energy products. The aggregated global renewable-energy scenarios made some years ago (Sørensen and Meibom 2000) maintained this type of solution by allowing large energy imports (mainly of liquid biofuels), and such a behavior will certainly be possible for a while in the future handling of the Japanese energy system. However, new emerging economies keep appearing with offers of less expensive consumer goods, and it is not certain that established economies such as the Japanese one can continue to assume a large trade surplus in the international market. It is therefore of interest to see if the strategic area of energy supply can be handled with nearly all resources derived from indigenous areas, which is thus the aim of the scenario to be constructed here. It is from the onset clear that the task is difficult and that some concessions will have to be made, e.g., related to the land and near-shore ocean areas to be set aside for energy production, but also in relation to the level of energy conversion and use efficiency to be implemented, which may have to be higher than the reference values considered in Chapter 4 for North America.

The 2050 electricity use assumed for Japan is 30% higher per capita than the one used in the reference scenarios of Chapter 4, but only 62% of the current use (reflecting efficiency improvements but not excluding new activities). The energy used in the Japanese transportation sector dropped 14% from 2000 to 2011 (IEA 2014) and is further assumed to drop by 25% to 2050, where average vehicles are assumed to be as efficient as the best today, and where many will use hydrogen. Per capita space heating in Japan is today modest, due of course to low-latitude climate but also to a fairly small per capita floor size. The 2050 scenario space heating demand shown in Figure 13.18 is half the reference value for the floor area of 60 m^2 per capita globally assumed in Chapter 4 (and presented in Figures 13.12 and 13.13), and would thus correspond to a heated area of 30 m^2 per capita.

Figures 13.19 and 13.20 enlarge the height profile and offshore wind resources from Figures 13.5 and 13.10. Figure 13.21 shows the wind-power resources that have to be tapped for the 2050 scenario to make Japan self-sufficient in sustainable energy supply. For offshore wind, the 0.1% of the potential to 50-m foundation depth is exploited. This fraction is the same as the one used in European and North American scenarios, and it is not believed that a much higher fraction

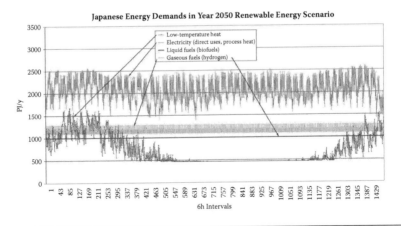

Figure 13.18 Time development over a year of the assumed Japanese energy demand in 2050. The electricity demand comprises uses in industry, commerce, and home, including space cooling. Transportation energy is divided between biofuels and hydrogen (as in the Chapter 4 scenarios), such that the available fuels derived from biomass residues are used, and the remaining mobile energy has to be furnished by conversion of electric energy to hydrogen (in reversely operated fuel cells, also called electrolyzers).

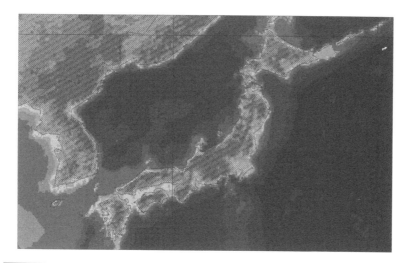

Figure 13.19 Height relief for Japan and Korea, to be used, e.g., for determining accessible offshore wind sites. The figure is a blowup of Figure 13.5 and uses the same shading definitions.

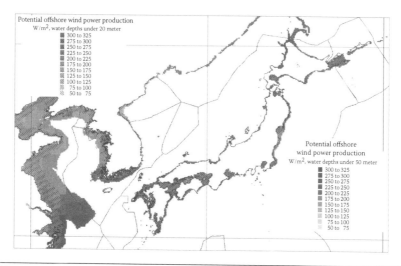

Figure 13.20 Offshore wind potentials in Korea–Japan region for foundation depths up to 20 m or 50 m. The figure is a blowup of Figure 13.10. The lines dividing territories at sea are the Exclusive Economic Zones (EEZ) defined by FAO (2014) to regulate marine fishing permissions.

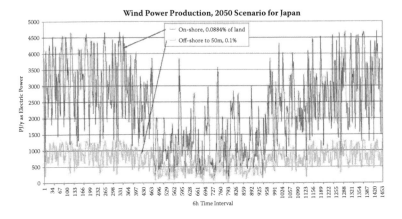

Figure 13.21   Time variation over a year of onshore and offshore wind production used in the Japanese energy scenario for 2050. The marked lower production during four summer months relative to the rest of the year is significant for the effort to match supply and demand. The scenario uses seasonal storage of hydrogen and fuel cells to handle the implied power intermittencies.

could be used because of turbine-shadowing effects (meaning that turbines "behind" others in the wind direction could receive winds of lower speeds if they are too close, because the energy in the wind would not have had time to replenish: it is known that wind speed profiles are reestablished by energy transfer from general circulation if 10–20 rotor diameters are allowed between turbines; Sørensen [2010]). For onshore wind, the same shadowing limit applies, but the reference wind turbine density used in Chapter 4 is much lower (swept area 0.01% of land surface area) due to considerations of other land usage. For Japan, it is found that a density of 0.0884% is needed in order to satisfy both the electricity demand and the need for hydrogen to vehicles and for smoothing of production variations.

The striking feature of Figure 13.21 is the persistent low winds during a four-month period in summer. Similar behavior was not present in the European and North American scenarios, and it places considerable pressure on the ability to handle the implied mismatch between demand and supply. As posited later in this discussion, photovoltaic power is not sufficient to help much, and the assessment of the intermittency problem suggests, in accordance with the conclusions previously made in Parts I to III of this book, that demand management is also insufficient due to the four-month-long displacement requirement. Since trade options were purposely left out (and currently there are no power-line connections between Japan and the continent), energy storage is the only workable solution. As hydrogen is already part of the scenario (for the transportation sector), the natural solution is to establish hydrogen stores. As mentioned in Chapters 8 and 9, these can be placed underground, with a low cost in cases where aquifers are present, as they are in some parts of Japan. The resulting variations over the year in filling of such hydrogen stores are shown in Figure 13.22. As the wind data pertains to a particular year (2000), one should allow for an amount of interannual variation. This will not be treated here, but deficiencies could be handled by generating a little additional electricity by slightly increased use of biofuels.

Figure 13.23 is an enlargement of the relevant part of the solar radiation shown in Figure 13.9, but using a finer grid. The time variation of power production using photovoltaic devices with an assumed 2050 efficiency of 20% is shown in Figure 13.24 for solar panels integrated into buildings and for central facilities placed on marginal land. The scenario assumes that the building-integrated solar collectors are of the PVT type, making use of the heat generated in the panel together with the assumed 18% of electricity. The recovered heat is taken as three times the power (typical of PVT systems [Sørensen 2000, 2002]). The performance of thermal solar collection depends sensitively on the inlet temperature of the fluid used (normally water or air), which is usually taken from a local heat store, the size of which relates to temperature. As mentioned earlier, a small store with high temperature gives the thermal part of the PVT collector a poor efficiency, while a large store gives higher efficiency but perhaps makes the fluid temperature too low for the intended application (space heating, hot-water production). The factor of 3 between heat and power generated assumes that the heat-store size is optimized relative to collector input and the building's heat takeout. The scenario assumes a collector area of 1 m^2 per capita, which is lower than the value used in scenarios for other parts of the world (Europe, North America) but probably adequate for Japan, due to the smaller floor area available to each person. Even if floor areas per capita increase further by 2050, this is likely to involve many high-rise buildings with less solar collection potential, due to shadowing effects and lower surface-to-volume ratios.

For the central photovoltaic power production, it should be noted that Japan has a relatively small area of marginal land available, that

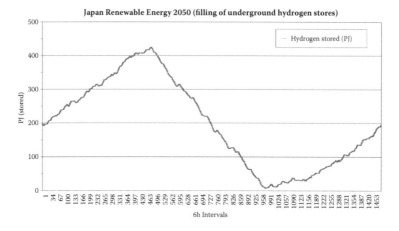

Figure 13.22 Time development of the level of filling over a year for the hydrogen storage facilities in the 2050 renewable energy scenario for Japan, reflecting the seasons with surplus wind energy for hydrogen production and the period with insufficient wind, where hydrogen from the stores is used to regenerate electricity. Along with this, a steady use of hydrogen in vehicles takes place over the entire year. The storage capacity corresponds to 70 days of average electricity demand.

furthermore is dominated by mountains and parks used for recreational purposes. Thus, only 0.16 percent of the land is set aside for solar power production, as compared to 0.2% for most of the European or North American scenarios. As a consequence, the Japanese will have to use more wind energy if they aim to be both sustainable and independent of energy imports. The scenario choice for onshore wind power (four to nine times the turbine density of Chapter 4 scenarios) need not create more competition for uses of marginal land, as wind turbines can be placed on agricultural land (Figure 13.17) without diminishing crop yields, because of the small physical footprint of the turbine towers.

The 2050 scenario arrives at the disposition of produced electricity during the year shown in Figure 13.25. The summer dip in wind power is partially compensated by use of the hydropower resources available in Japan. However, these are limited, and also electricity from hydrogen-fueled fuel cells needs to be invoked, requiring in turn that more wind power be produced during the eight months with high winds. This determines the total amount of wind energy resources to employ, comprising, as mentioned, both the offshore options (which are present quite close to the Japanese coast due to the steep descent of the ocean floor) and use of the many good land sites to an extent of

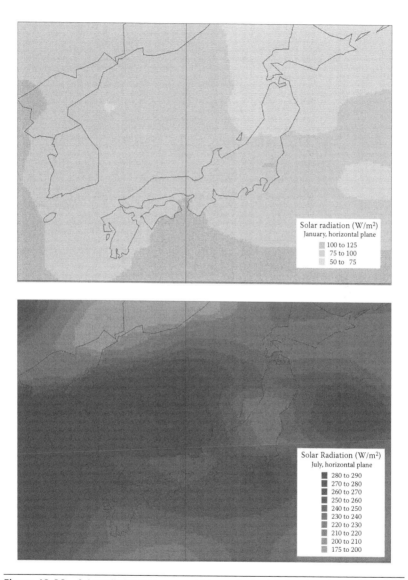

Figure 13.23   Solar radiation in January and July on a horizontal plane in and around Japan and Korea. Similar to Figure 13.9, except calculated with use of 0.125° × 0.125° radiation data for 2000 provided by ECMWF (2014).

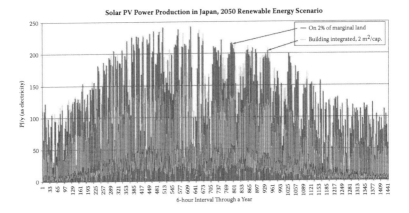

Figure 13.24 Time variation over a year of building-integrated and marginal-land photovoltaic power production used in the Japanese energy scenario for 2050. Note a weakening of the production during some summer periods. The scenario uses 0.16% of the available solar resources on marginal land, slightly less than in the North American scenarios of Chapter 4, due to competing, mainly recreational, land uses. The building-integrated solar collector area, on the other hand, is taken as fairly modest (1 m² per capita) due to the lower per capita floor area assumed available to the Japanese population, relative to the situation in North America.

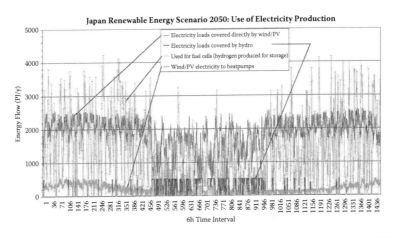

Figure 13.25 Time development of the disposition of electricity production in the 2050 renewable energy scenario for Japan. Hydroelectricity is employed in an attempt to make up for the reduction in wind energy during the summer months. Additional electricity has to be produced by use of stored hydrogen in fuel cells.

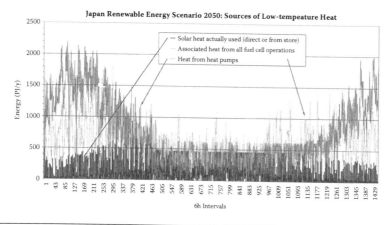

Figure 13.26 Time development of the available sources of low-temperature heat in the 2050 renewable energy scenario for Japan. The waste heat from fuel-cell operations (forward or backward) is used where district heating pipelines from the sites of hydrogen storage to the individual users are or can be economically established.

an area swept that is 0.0884% of the land area. As Figure 13.25 also shows, the wind surplus during the eight months of high winds allows substantial amounts of hydrogen to be produced, but not continuously. The reverse fuel cells (electrolyzers) have to be operated when wind exceeds electricity demand, which is often just during part of the day or at one- or two-day intervals. Finally, electricity covers some low-temperature heat demand missed by the solar thermal systems through the use of heat pumps. The amounts naturally peak during winter, as shown in Figure 13.25. Figure 13.22 showed the filling of the hydrogen stores mentioned previously, depicting the dramatic summer period of rapidly emptying the stores.

Figure 13.26 shows the time sequence of making use of the heat production in a way similar to Figure 13.25 for electricity. Hot-water supply is furnished by the rooftop solar PVT systems, by heat pumps, and by heat from fuel-cell operations. Much of the modest demand for space heating during spring to autumn is solar energy. During winter, demand is higher, and heat from heat pumps now exceeds that derived from the PVT collectors and fuel-cell "waste heat." Because the heat from fuel-cell operations has to be transmitted to the load locations, only areas where district heating lines exist (or can be established) can make use of this option, unless fuel-cell operation is decentralized as suggested in Sørensen (2012a).

Table 13.2   Summary of 2050 Scenario for Japan

UNIT: PJ/Y	LOW-T HEAT	ELECTRICITY	GASEOUS FUELS	LIQUID FUELS
Delivered energy demand	718	2087	1146	1000
Onshore wind-power production		2319		
Offshore wind-power production		707		
Hydropower production		509		
Building-integrated PV production		108		
Marginal land PV production		28		
Biofuels from agricultural residues				405
Biofuels from forestry residues				153
Biofuels from aquaculture				450
Solar thermal energy produced	300			
Electricity for dedicated uses		1997		
Electricity for hydrogen production		1576	1418	
Electricity to heat by heat pumps	513	171		
Liquid biofuels for transportation				1000
Hydrogen for use in vehicles			1146	
Hydrogen for electricity production	(>82)	163	272	
Heat from fuel cell operations used	73			
Solar thermal heat used directly	127			
Low-temperature heat from stores	5			
Discarded or lost solar heat	168			
Import and export		0	0	0

Note: The assumed 2050 population is 110 million.

Table 13.2 summarizes the 100% renewable energy Japanese 2050 scenario in a way similar to the tables presented in Chapter 4 for North America and in Chapter 12 for Denmark. It is evident that some of the aggressive uses of renewable resources could be avoided if the intermittency problems were handled by energy exchange with other countries. Indeed, even collaboration within Japan is extremely important, and the scenario assumes that electricity transmission within Japan can be performed as needed. The current splitting of power supply over a number of private companies with low capacity for transfers and using two different alternating current cycles (cf. Figure 5.8) has to be remedied, as the aftermath of the Fukushima nuclear accident has already signaled. Regarding international connections, South Korea is the natural first choice, but as the following section will show, the utility of this solution is limited due to the renewable energy situation in Korea being quite similar to that of Japan. Collaboration with its other relatively close neighbor, Siberian Russia, offers more opportunities, but is currently not favored politically. Collaboration with China will likely be of mutual benefit, but for electricity transmission, the problem is that the natural path of connection would be through North Korea, again with political obstacles presently seen as very substantial.

13.2 A Korean Renewable Energy Future

The South Korean situation in many ways resembles the Japanese one: High population density and making up for not being self-sufficient in food and energy by having created a trade surplus of industrial products for general consumers. This strategy started later than in Japan, primarily due to the negative effects of the civil war with North Korea, which became a testing ground for superpower interventions as part of the subsequent series of clashes aimed to delineate the imperialistic aspirations of the post–World War II era. The scenario for 2050 in consequence has many similarities to the Japanese one described in Section 13.1. Figure 13.27 gives the variations in energy demands of South Korea over the year. Per capita electricity demands are 30% over the reference case of Chapter 4, but still only 68% of today's

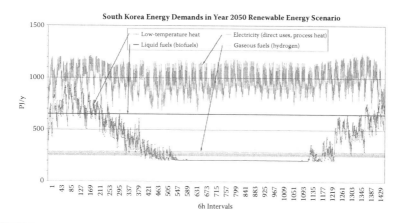

Figure 13.27 Time development over a year of the assumed South Korean energy demand in 2050. The electricity demand comprises uses in industry, commerce, and home, including space cooling. Transportation energy is divided between biofuels and hydrogen in such a way that the available fuels derived from biomass residues are used and the remaining mobile energy has to be furnished by conversion of electric energy to hydrogen.

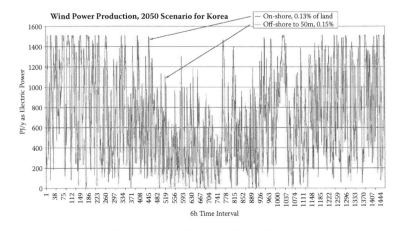

Figure 13.28 Time variation over a year of onshore and offshore wind production used in the South Korean energy scenario for 2050. Very substantial offshore (swept area 0.15 times sea surface area) and onshore resources (swept area 0.13 times land surface area) are exploited in the scenario. As in Japan, there is a significantly lower wind power production during four summer months relative to the rest of the year. The scenario uses seasonal storage of hydrogen and fuel cells to handle the implied power intermittencies.

due to efficiency improvements. Space-heating demands are, as for Japan, taken as the reference value for 30 m² per capita, considering the high population density and the wide scope for improving the building standards by proper insulation. The transportation energy demand has been unchanged from 2000 to 2012 (some 1100 PJ/y) and is assumed in the sustainable scenario through a combination of efficiency improvements and activity increase to decline to 910 PJ/y.

The height relief and offshore wind map in Figures 13.19 and 13.20, respectively, show that sufficiently shallow waters are present along the west coast of Korea, but with lower wind speeds than around Japan. The time variations of South Korean wind resources shown in Figure 13.28 assume an offshore turbine density of 0.15% swept-area to sea-surface ratio (increased relative to the case of Japan due to the lower wind speeds off Korea), and the larger production on land is accomplished by sweeping 0.13% of the land area, again higher than the value used for Japan. However, this is necessary to make the 100% renewable scenario work, as the power derived from photovoltaics (Figures 13.22 and 13.29) is also quite limited, despite the 2 m² per capita set aside for solar panels on buildings (two times more than in Japan, but the same as was used in the Chapter 4 scenarios), and very little coming from marginal land, even at 10% utilization. The

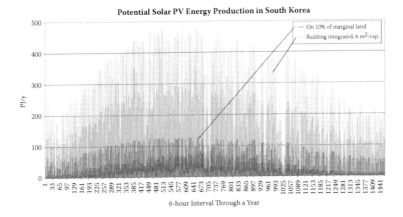

Figure 13.29 Time variation over a year of building-integrated and marginal-land photovoltaic power production used in the South Korean energy scenario for 2050. Because very little land is designated as marginal in Korea, the 0.2% used for PV contributes almost nothing. For building-integrated PVT collectors, an area of 2 m² per capita is assumed, as in the US scenario in Chapter 4.

reason is that land designated as marginal is minute in South Korea, with agricultural land and forests making up most of the land area.

The time development of the uses to which the produced electricity is supplied is depicted in Figure 13.30, showing that hydropower is insignificant and that power converted to heat during winter by heat pumps is as prevalent as in the Japanese scenario, but that coverage

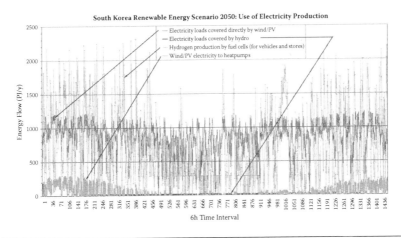

Figure 13.30 Time development of the disposition of electricity production in the 2050 renewable energy scenario for South Korea. Hydroelectricity is insignificant and heat-pump production of heat peaks during winter as in the Japanese scenario. The summer dip in wind power availability has been compensated for by dedicating a larger area to wind, so that only the capability of producing hydrogen is negatively affected during the four summer months, not the dedicated electricity uses.

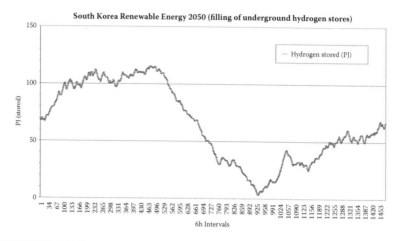

Figure 13.31   Time development of the level of filling over a year for the hydrogen storage facilities in the 2050 renewable energy scenario for South Korea, similar to the Japanese case but with larger excursions from the average trend. Along with energy-storage uses of hydrogen, a steady use in vehicles is taking place over the entire year. The storage capacity corresponds to 45 days of average electricity load.

of direct electricity needs is achievable year-round by wind and solar power, or by drawing from stored energy. As in Japan, intermittency is taken care of by the use of underground hydrogen stores, with the pronounced dip in filling level during summer shown in Figure 13.31 being a little less pronounced than for Japan (Figure 13.22). Building up the level of stored hydrogen is, as in Japan, achieved from intermittent surplus power produced primarily during the eight nonsummer months.

As regards low-temperature heat, the time development of coverage in the 2050 scenario is shown in Figure 13.32. Winter space-heating demand is higher than in Japan, but also here covered by a mix of solar thermal energy (from combined PVT panels, as there is hardly sufficient south-facing building surfaces to allow separate power and thermal collectors) in combination with heat pumps. Again, the waste heat from fuel-cell operation is occasionally used in places where heating lines are in place.

The 2050 renewable-energy scenario for South Korea is summarized in Table 13.3. In South Korea as well as in Japan, one may ask whether the high density of wind turbines on land, as well as the use of solar cell panels, will be acceptable to the population. As already stated, the simplest alternative would be energy import, with the associated

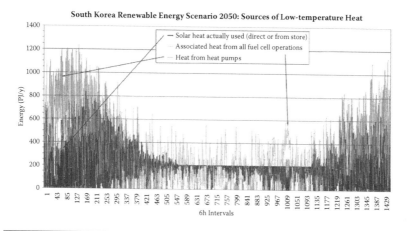

Figure 13.32   Time development of the available sources of low-temperature heat in the 2050 renewable energy scenario for South Korea.

Table 13.3   Summary of 2050 Scenario for South Korea

UNIT: PJ/Y	LOW-T HEAT	ELECTRICITY	GASEOUS FUELS	LIQUID FUELS
Delivered energy demand	383	991	260	650
Onshore wind-power production		712		
Offshore wind-power production		625		
Hydropower production		23		
Building-integrated PV production		117		
Marginal land PV production		4		
Biofuels from agricultural residues				232
Biofuels from forestry residues				326
Biofuels from aquaculture				95
Solar thermal energy produced	300			
Electricity for dedicated uses		875		
Electricity for hydrogen production		618	556	
Electricity to heat by heat pumps	150	50		
Liquid biofuels for transportation				650
Hydrogen for use in vehicles			260	
Hydrogen for electricity production	(>89)	178	297	
Heat from fuel cell operations used	47			
Solar thermal heat used directly	154			
Low-temperature heat from stores	27			
Discarded or lost solar heat	119			
Import and export		0	0	0

Note: The assumed 2050 population is 52 million.

requirement of maintaining a large trade surplus against the rest of the world well into the future. Seen from an environmental point of view, the high density of wind and solar conversion equipment is not a problem, because if better solutions should be found some time in the future, both wind turbines and solar collectors can be removed without leaving any impact on the rural or the urban environment.

13.3　A Chinese Renewable Energy Future

The Chinese situation is quite different from that of Korea or Japan, because China covers land areas of several different climates due to the span of latitudes, of coastal and continental contexts, and of height above sea level (see Figure 13.5). This affects the potential for production of energy from the different renewable energy sources, and it affects demand due to the variations of both climatic conditions and population density (Figure 13.2), characterized by high values in the coastal and eastern regions of high urbanization, but low values in the high-altitude plains in the west of the country. This means that energy installations such as wind farms, centralized photovoltaic plants, and perhaps even growth of crops suitable for poor soil quality and water shortage may be placed on the vast marginal land areas in the west, where they could take up a fairly high fraction of the land area without interfering with other uses. For biomass production, artificial irrigation would make a large difference (see Figure 13.6). The marginal land area in western China is very substantial in size, as shown in Figure 13.16, and current agriculture is concentrated in the east and south of China (Figure 13.17). The need for making use of substantial land areas for wind is underlined by the fairly modest but still quite useable wind speeds available on the high plains (Figure 13.7). The solar resources on the high plains are excellent (Figure 13.9). A drawback for both solar and wind energy from the northwest region is, of course, that the energy has to be transferred to the main load areas to the east. This would imply establishment of considerable amounts of new capacity of electricity transmission lines between west and east, at a substantial cost. At present, such lines are few, as seen in Figure 5.8. However, an emphasis on long-range power transmission was realized a little farther north, in the former Soviet Union, a long time ago (see Figures 5.7 and 5.8).

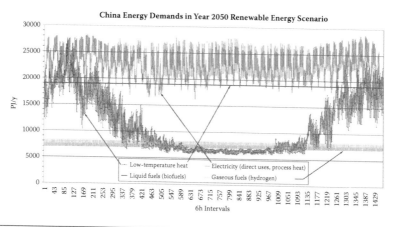

Figure 13.33 Time development over a year of the assumed Chinese energy demand by 2050. The electricity demand comprises uses in industry, commerce, and home, including space cooling. Transportation energy is divided between biofuels and hydrogen in such a way that the available fuels derived from biomass residues are used, and the remaining mobile energy has to be furnished by conversion of electric energy to hydrogen.

Chinese energy demands assumed in the 2050 scenario are displayed in Figure 13.33 as they develop in time over a year. Because of the northern continental climate regions, Chinese space-heating demand is substantially higher than in Korea. As China can be self-sufficient in food, agricultural residues are fairly large and allow, together with forestry residues, coverage of more than 70% of the transportation demands. The rest is in the scenario covered by hydrogen, which, as discussed later in this section, is made available in substantial amounts for this scenario. The scenario wind-power production is shown in Figure 13.34. The offshore production off the southern shores (cf. Figure 13.10) for a swept-to-surface-area ratio of 0.15 is larger than that of South Korea and is similar to the one used in the Japanese scenario, although produced over a larger area with wind speeds lower than those of the Japanese offshore regions. However, relative to the total Chinese population, the offshore power production is small, and much larger quantities can be and are produced from land areas, and notably with only 0.075% of the entire land area swept, at the locations indicated in Figure 13.7, predominantly situated in the western plains, but with smaller amounts along the south coast and in Inner Mongolia.

The power production by photovoltaic panels is shown in Figure 13.35. It is dominated by centralized plants and is taking up 0.31% of marginal land, again with the western high plains

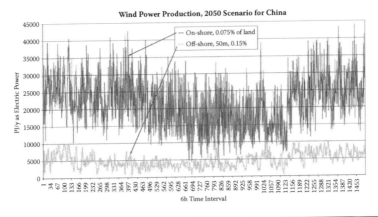

Figure 13.34   Time variation over a year of onshore and offshore wind production used in the Chinese energy scenario for 2050. The variations over the year are somewhat unsystematic and reflect the different conditions in different parts of China. The summer decline is smaller than in Japan and Korea and stretches longer into the autumn period.

contributing most and conceivably being able to accommodate even more solar plants. Increasing the building-integrated per capita solar collection area from the assumed 1 m^2 per capita (considering low roof areas and room for heat stores available in cities) within reasonable limits will not provide sufficient amounts of power or heat, implying that low-temperature heat requirements will have to be met by a combination of heat pumps, solar and waste heat.

Figure 13.36 shows the disposition over the year of electric power produced by the sources mentioned. About 94% of the dedicated

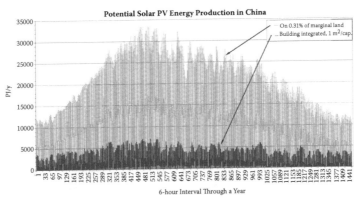

Figure 13.35   Time variation over a year of building-integrated and marginal-land photovoltaic power production used in the Chinese energy scenario for 2050. The production on 0.31% of the marginal land, mostly in the northwest, is much larger than what can be derived from the 1 m^2 per capita roof areas assumed on average used for PVT panels by the 2050 Chinese citizens.

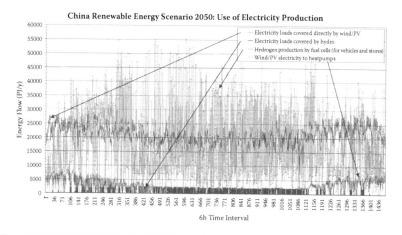

Figure 13.36 Time development of the disposition of electricity production in the 2050 renewable energy scenario for China. Hydroelectricity is only slightly expanded relative to the present level. The late summer dip in wind power availability has been compensated by dedicating a larger area to wind, so that variability only affects hydrogen production (see Figure 13.37 below).

electricity demand can be provided directly from the solar, wind, and hydro installations in the scenario. The rest comes from hydrogen stored for later regeneration of electricity at times of insufficient direct production. As mentioned, electricity delivered to heat pumps plays a decisive role in providing low-temperature heat needs, and the remaining electricity production is used to produce hydrogen. Some 30% of the transportation sector demand is covered by hydrogen, the rest being serviced by biofuels used in internal combustion engines. Figure 13.37 shows the level of filling for the underground hydrogen stores, explaining why the hydrogen production has to be larger than the sum of demands. The hydrogen filling level is getting low two times during the year, in February–March and again in September–October. If hydrogen production were lower, it would not be possible to satisfy demands during these periods. The implied surplus energy at times where no national demand requires it will be available for export (see Table 13.4; the scenario assumptions cause the potential export to be in the form of electricity). One should note that the store filling develops quite differently over the year than that in South Korea (Figure 13.31) or Japan (Figure 13.22). There is no four-month-long period of declining energy in the hydrogen stores of the Chinese system, which are therefore more modest in total capacity than the South Korean or Japanese ones, considering the differences in population

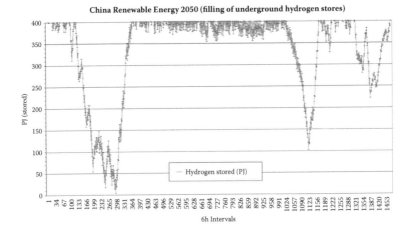

China Renewable Energy 2050 (filling of underground hydrogen stores)

6h Intervals

Figure 13.37   Time development of the level of filling over a year for the hydrogen storage facilities in the 2050 renewable energy scenario for China. Note that the behavior is quite different from the situation in South Korea and Japan. A substantial amount of surplus hydrogen is produced over the year (and it could be larger if the storage capacity were larger). It may be available for export, or the electricity could be exported before being converted to hydrogen, but the time-sequence of generating power for export is quite intermittent (see Figure 13.36). Along with energy-storage uses of hydrogen, a steady use in vehicles takes place over the entire year. The capacity of the hydrogen stores is just 6.3 days of average electricity demand.

sizes. Figure 13.37 shows that the Chinese stores are full (hitting the ceiling assumed for the storage capacity) during extended periods of the year, implying that more hydrogen could be produced if the stores had been larger. The result would be that the hydrogen filling at the end of the year could exceed that at the beginning by about a factor of three. Evidently, this would not be manageable over several years, so the implication is that China, if desired, could export large quantities of hydrogen rather than the potential electricity export selected in the scenario. (More hydrogen production would, of course, diminish the electricity surplus correspondingly.) The scenario considers electric transmission lines as cheaper than hydrogen pipelines and has therefore chosen to keep the surplus as electricity (Table 13.4). The same argument was invoked in the North American scenarios of Chapter 4.

For low-temperature heat, Figure 13.38 shows the small solar contribution and the large heat-pump contribution. As indicated in Table 13.4, there is additional solar PVT thermal energy available, but limits to urban storage capacity are considered to make utilization difficult. Waste heat from the fuel-cell operations is used at irregular periods throughout the year, in cases where cities reasonably close to

Table 13.4 Summary of 2050 Scenario for China

UNIT: PJ/Y	LOW-T HEAT	ELECTRICITY	GASEOUS FUELS	LIQUID FUELS
Delivered energy demand	11,561	23,123	7293	19,003
Onshore wind-power production		21,633		
Offshore wind-power production		4799		
Hydropower production		2043		
Building-integrated PV production		3054		
Marginal land PV production		10,287		
Biofuels from agricultural residues				11,814
Biofuels from forestry residues				6408
Biofuels from aquaculture				853
Solar thermal energy produced	6005			
Electricity for dedicated uses		21,700		
Electricity for hydrogen production		17,883	16,095	
Electricity to heat by heat pumps	8049	2683		
Liquid biofuels for transportation				19,003
Hydrogen for use in vehicles			7289	
Hydrogen for electricity production	(2713)	5427	9045	
Heat from fuel cell operations used	2407			
Solar thermal heat used directly	805			
Discarded or lost solar heat	5200			
Import requirement		0	0	0
Export potential		2974	0	0

Note: The assumed 2050 population is 1516 million.

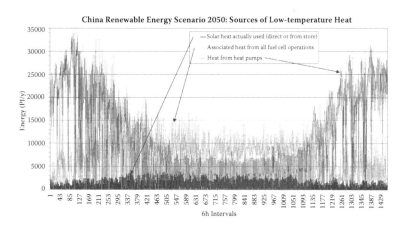

Figure 13.38 Time development of the available sources of low-temperature heat in the 2050 renewable energy scenario for China.

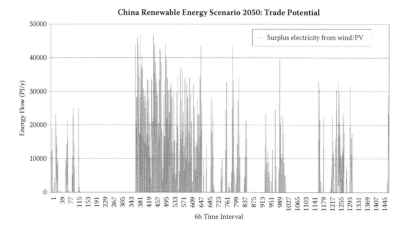

Figure 13.39   Time-development of surplus electricity available for export in the 2050 renewable energy scenario for China.

the fuel-cell plants can be supplied through district heating lines (of which a fair amount are already present).

Figure 13.39 shows the time distribution of China's electric power export potential in 2050. It is concentrated on particular periods of the year and is largest in the spring months. Gaps have durations in excess of two months, which may not be acceptable to potential importing countries, although in the case of Korea and Japan, the distribution is fairly convenient. Still, and particularly if other export countries are selected, keeping the exports as hydrogen may be the best solution. The scenario is summarized in Table 13.4.

13.4 Conclusion

The three scenarios considered in this chapter indicate that while Japan and South Korea have to stretch resource usage far to realize a 100% renewable energy supply system, the conditions are more favorable in China, which even with a more modest resource utilization is in a situation that allows energy export. Unfortunately, quite long power transmission lines will be needed to reach South Korea or Japan, at least if passage through North Korea is excluded for reasons of supply security in the importing countries. This does not mean that the Chinese scenario is without challenges, particularly because of the long distances by which energy has to be transferred due to mismatch between the locations of energy demand and production.

References

Claus, S., N. Hauwere, B. Vanhoorne, F. Hernandez, and J. Mees. 2014. Exclusive economic zones boundaries. In *World maritime boundaries, v6*. Flanders Marine Institute. http://www.marineregions.org/downloads.php.

Columbia University. 2013. Data library. http://iridl.ldeo.columbia.edu.

ECMWF. 2014. *40 year re-analysis data server data sets*. European Centre for Medium-Range Weather Forecasts. http://apps.ecmwf.int/datasets/.

FAO. 2014. *Fishery zones*. Food and Agriculture Organization of the United Nations. (May be downloaded from Claus et al. [2014].)

IEA. 2014. *World energy statistics 2013*. Paris: International Energy Agency/ OECD.

Melillo, J., A. McGuire, D. Kicklighter, B. Moore, C. Vorosmarty, and A. Schloss. 1993. Global climate change and terrestrial net primary production. *Nature* 363:234–40.

NCAR. 2006. *QSCAT/NCEP blended ocean winds ds744.4*. US National Center for Atmospheric Research, Data Support Section. http://rda.ucar.edu/datasets/ds744.4.

NOAA. 2004. *ETOPO5 elevation dataset*. US National Geophysical Data Center. http://iridl.ldeo.columbia.edu.

Smith, A. 1776/1904/1976. *An inquiry into the nature and causes of the wealth of nations*. Repr. London: Methuen & Co.; Chicago: University of Chicago Press. http://www.econlib.org/library/Smith/smWN.html.

Sørensen, B. 1982. Energy choices: Optimal path between efficiency and cost. In *Energy resources and environment*, ed. S. Yuan, 279–86. New York: Pergamon Press.

Sørensen, B. 2000. PV power and heat production, an added value. In *Proc. 16th European Photovoltaic Solar Energy Conference*. Vol. 1, 1848–51. London: James & James.

Sørensen, B. 2002. Modeling of hybrid PV-thermal systems. In *Proc. 17th European Photovoltaic Solar Energy Conference*. Vol. 3, 2531–38. Florence, Italy: WIP-ETA.

Sørensen, B. 2008a. A sustainable energy future: Construction of demand and renewable energy supply scenarios. *Int. J. Energy Research* 32:436–70.

Sørensen, B. 2008b. A new method for estimating off-shore wind potentials. *Int. J. Green Energy* 5:139–47.

Sørensen, B. 2010. *Renewable energy: Physics, engineering, environmental impacts, economics and planning*. Burlington, MA: Academic Press/Elsevier.

Sørensen, B. 2011a. *Life-cycle analysis of energy systems: From methodology to applications*. Cambridge, UK: RSC Publishing.

Sørensen, B. 2011b. Mapping potential renewable energy resources in the Mediterranean region. In *Recent developments in energy and environmental research*, ed. E. Malevi, 23–35 (chap. 3). Athens: Athens Institute for Education and Research.

Sørensen, B. 2012a. *Hydrogen and fuel cells. Emerging technologies and applications*. Burlington, MA: Academic Press/Elsevier.

Sørensen, B. 2012b. *A history of energy*. New York: Earthscan/Routledge.

Sørensen, B. 2014a. Conditions for a 100% renewable energy supply system in Japan and South Korea. Article in preparation.

Sørensen, B. 2014b. Feasibility of a 100% renewable energy supply system for China. Article in preparation.

Sørensen, B., and P. Meibom. 2000. A global renewable energy scenario. *Int. J. Global Energy Issues* 13:196–276.

UN. 2013. *World population policies 2013*. Dept. Economic and Social Affairs, Population Division, Report ST/ESA/SER.A/341, New York.

Ushiyama, I., H. Nagai, T. Saito, and F. Watanabe. 2010. Japan's onshore and offshore wind energy potential as well as long-term installation goal and its roadmap by the year 2050. *Wind Engineering* 34 (6): 701–20.

WEC. 1995. *Survey of energy resources*. London: World Energy Council.

Weizsäcker, E. von, A. Lovins, and L. Lovins. 1997. *Factor 4: Doubling wealth—Halving resource use*. Sydney: Allen & Unwin.

Weizsäcker, E. von, C. Hargroves, M. Smith, C. Desha, and P. Stasinopoulus. 2009. *Factor 5: Transforming the global economy through 80% improvements in resource productivity*. London: Earthscan.

14
FINAL REMARKS

Looking broadly at intermittency in energy supply systems, this study has found that examples of intermittency, disruption, and variation in the ability to match demand and supply can be found in all existing and contemplated energy systems. However, the characteristics of intermittency situations can be quite different for different types of systems. Those systems with supply from a few large installations are vulnerable due to risk of failure, whether technical or caused by external circumstances, and so are energy distribution systems based on a few large channels of transport (e.g., power lines and pipelines). This also goes for supply based on import of energy from a small number of suppliers, including inland imports and particularly if involving international import from unstable regions. In contrast, smaller, decentralized, or modular energy supply units inherently reduce the risk of some avenues for creating intermittency situations, but if they are based on renewable energy flows, the ability to regulate supply often becomes limited, so that intermittency situations become much more probable.

The possibility of having intermittency in the supply–demand matching is one thing, but it has to be juxtaposed with an assessment of the options for dealing with the intermittency. In the case of accidents in large installations, there are rarely direct possibilities for avoiding at least short-term intermittency caused by associated power disruptions, except when substantial backup has already been established. Such "doubling" of capacity is rarely established to the fullest extent, but there may be partial backup solutions, e.g., through agreements with neighboring countries (or regions within extended countries). This shifts the expense from building duplicate energy production plants to building transmission connections with more capacity than needed for daily operation. In today's world, there are many regions where the establishment of transmission facilities for intermittency emergencies has not been undertaken, and there are a few regions where they are already in place.

For the systems with intermittent source influx, such as those based on renewable energy, a number of available options have been surveyed, such as use of energy storage or demand manipulation. The trade between regions or countries may also be invoked here, because the time development of demand and supply from the available sources may have different profiles in different regions, and these may allow adaptation by energy interchange to the situations where source intermittency would otherwise have disrupted supply. The tools in this type of transfer are power transmission lines and fuel pipelines as well as other means of transport. The challenge is that electricity can only be stored indirectly (in altered forms), in contrast to many fuels that can be stored directly and therefore can be used to handle source deficiencies directly at a lower cost, without conversion from energy stored in some other form.

This book has provided several examples of using these methods for various energy systems. In Chapter 4, a number of reference energy-system models were constructed for the North American countries based on the use of available renewable energy resources. It turned out that because of the use of intermittent resources for both time-urgent power and heat demands (and for demands that could be satisfied by stored energy from hydro reservoirs or hydrogen stores), then most of the time there were enough direct source flows to cover the dedicated uses of electricity and heat. The irregular occurrence of surpluses would then be used for hydrogen production or for holding back some water in hydro reservoirs for later use, and only a modest need was identified for regenerating power from stored hydrogen or for drawing heat from heat stores. Despite this positive functioning of the systems with 100% renewable energy, where a large portion of the sources are inherently intermittent, Chapter 5 showed that the systems could function better if transmission between the countries were substantially increased, suggesting that electric power transmission would be less expensive than hydrogen or biofuel transport through pipelines (Chapter 6). The option of transporting fuels by ship (Chapter 7) is in effect at present for fossil fuels, but this is expected to play a lesser role in renewable energy systems. The use of energy storage (Chapters 8–10) was already invoked in the scenarios of Chapter 4, notably in the case of storing hydrogen underground in salt domes, aquifers, or similar structures. Of course, the role of hydrogen in future energy systems can be, and currently is, debated, but it is a convenient storage medium because of the possibility of avoiding pollution in a subsequent conversion to other energy forms. Even in the case where fuel cells fail to deliver their promise, hydrogen storage would still be an interesting option, although regeneration of electric power would have to be done by internal combustion, likely in gas turbines. Heat storage is expensive and inconvenient in many cases, and the scenarios avoid this by having only modest heat stores in the building-integrated solar energy collection systems, instead relying on conversion of electric energy by use of heat pumps (with air or soil as low-temperature reservoirs for urban settings, supplemented by rivers or lakes in central installations for areas with district heating lines available).

The proposed use of load management for dealing with intermittency (Chapters 11 and 12) is in many ways disappointing, according to the simulation studies for Denmark presented. The time deferment acceptable for different types of loads is limited, often to hours or at most days, which turns out to have a small overlap with the requirements set by the intermittency of renewable sources such as wind energy, for which only demand shifting of one, two, or three weeks would have the full desired effect. For solar energy, such as photovoltaic power, load management can help by shifting loads from dark hours to midday, but the effect here also is modest, due to the prevalence in many places of rain or cloudy weather during extended (weeklong) periods throughout the year.

The South-East Asian scenarios worked out in Chapter 13 support ost of the remarks made above. Due to the small area of marginal land available in Korea and Japan, the scenarios for these countries require quite high acceptance of solar panels in urban areas and quite high density of wind turbines at all suitable sites. The scenario is further stressed by the four-months long summer period of low winds, in a period with high electricity demands. Although the scenario system simulation shows that the system can depend on renewable energy alone, also with consideration of the differences in resource inflow between years, there remains an inviting alternative of importing some energy, continuing the present policy of creating an industrial excellence allowing export earnings to exceed industrial import payments, and thereby leaving room for import of food and energy. The question is of course if the global development jeopardizes this policy, but it is likely not the case as long as there are only few regions needing to pursue this path. In other words, the situation of Japan and South Korea, with high population density and fairly limited renewable resources, but only a small percentage of the world's population, should not prevail in large parts of the World. However, this could well happen, if population stability is not forthcoming. The United Nations' projection assumes that the entire world will be on a development track toward prosperity by 2050, and that population will therefore stabilize. Unfortunately, this is hardly realistic, considering the present development characterized by corruption and conflict, notably in Africa, but also in countries like India, that have given up family planning and instead work toward creating a modest-size elite of rich people not sharing their wealth with the rest of the population. This is not the recipe for a stable society, and if countries the size of India or China are pursuing a route continuing to depend on having a large trade surplus, they will need a huge underpaid segment of society, comprising as a minimum industrial workers. It is difficult in this light to see a development toward a global sustainable energy future.

China is doing quite well in the scenario sketched in Chapter 13. This is in contrast to one earlier global scenario, which was lumping each continent together as one region and found that the Asian region needed major energy imports from countries with a surplus, e.g., Europe, North America, and South America (Sørensen and Meibom 2000). The present scenario uses updated resource data, including a much improved model for offshore wind sources, and in contrast to the earlier model of annual means, the matching between supply and demand is now made with realistic time steps of six hours. The outcome is that China can in fact cover its carefully optimized energy demands estimated for 2050, without putting undue strain on the commitment of areas for wind turbines or photovoltaic arrays. This is very positive because, as noted previously, the import option is problematic for countries with a large population such as China. The current practice of the Chinese government is to shop around for mineral resources in Africa and in the Arctic, which clearly does not offer a sustainable long-term option.

The case studies for North America (Chapter 5) showed the importance of having sufficient transmission capacity available for developing viable renewable energy scenarios. By transmitting electric power from Canada to the United States and from the United States to Mexico, followed by local conversion to hydrogen, it was possible to satisfy transportation needs beyond what local biofuels could deal with. It was suggested that expansion of transmission capacity between these three countries should have a high priority if future sustainable energy systems are to work smoothly and, in the short term, to avoid blackouts and other energy disruptions.

In Europe, transmission capacity is already much higher, and plans for continued upgrading are being implemented. However, for the scenarios of Southeast Asia, the time-simulations performed did not support the suggestion that energy transfers between Japan and South Korea via undersea electric transmission cables would lead to a more resilient renewable energy system. The main obstacle was the coincidence of a prolonged low-wind period in both countries. China, on the other hand, was found to possess a power surplus available for export, but in the present political climate, this option may be seen as contrary to supply security requirements, at least from the point of view of South Korea and Japan. One trade option would call for an underwater cable from China to South Korea, bypassing North Korea, but direct energy dependence on China may seem undesirable to South Korea and Japan, given the several sudden changes in the Chinese power structure and its international behavior over the preceding 60–70 years.

Similarly, a fairly obvious solution, pointed out in Chapter 13, would be a power connection between Northern Japan and the eastern Russian Siberia, but again, political problems caused by territorial disputes seem to rule out such collaboration at the present time. Although no scenario for Siberian Russia has been worked out here, sizeable wind resources can be seen from Table 13.1 and Figure 3.4 to be present at northern and eastern Russian shores, in addition to hydro power (Figure 13.8), and since the indigenous population is very small (Figure 13.2), the energy demand for power and even for space heating (see Figures 13.12 and 13.13) is small enough to allow a substantial amount of export to Japan. One may hope that the political tensions are dissolved before the time where the change to a new energy system becomes a necessity.

The current scenarios actually do what energy scenarios are supposed to do: They show that certain energy systems are inherently consistent and can do the job, but they do not show how to get from here to there. This requires an implementation analysis, which has not been attempted in this book. The basic definition of scenarios— exploring a few out of a large (infinite?) number of possible snapshots of future situations and then establishing their consistency (Sørensen 2000; Petersen et al. 2001)—has to be supplemented by an implementation analysis, identifying the kind of political and industrial decision that have to be made in order to accomplish the transition from the present system to the scenario system. (A brief example may be found in Petersen et al. [2001].)

Among the dimensions that should be central in any energy system assessment is the environmental one. A number of different energy technologies have been explored in this book, considering their roles in mitigation of any negative effects of intermittency in the ability to satisfy energy demands. An evaluation of the overall suitability of alternative system components must naturally cover many aspects in addition to that of intermittency. Important among these are the impacts that the system can have on the environment, that is, the global environment (e.g., climate issues), the local environment (e.g., air pollution), and the human environment (such as health and work conditions). Negative environmental impacts are evident in some cases, such as nuclear energy with its periodic catastrophic accidents or oil shale extraction with its injection of potentially harmful substances into soils, but the entire set of impacts is often poorly known at the time of implementation, and different impacts are often incommensurable, making it difficult or impossible to conduct economic comparisons between the available options. The scientific background for such assessments has recently been surveyed by Sørensen (2011) using the life-cycle analysis and assessment method to cast some light on the help that decision makers can get in arriving at their choices. The assessment is, in many cases, quite uncertain due to the dependence of the impacts on the entire social entourage. For example, biofuels derived from agricultural residues would have lesser impacts than biofuels based on edible biomass (cereal grains, corn, etc.) if the magnitude of the world population demands including consideration of global food availability. Both types of biofuels have air pollution impacts when combusted, and both have long-range impacts on soil quality in the likely case that nutrients are not recycled (by extracting them from the biofuel production process and transporting them back to the fields). However, large differences in valuation may occur as a result of the economic rules of play in a society using biofuels: Unfettered market economies would make companies start with the cheaper biofuel production based on food biomass and only move on to residues, requiring a more expensive biochemical degradation process, when the market has driven food prices up to a level making the extra expense of using residues as feedstock economically advantageous. A quite different situation would occur if legislation were put in place from the start, restricting the free market from playing with food as energy feedstock and plainly requiring nutrient recycling. Some countries otherwise adhering to the market economy paradigm have introduced some such legislative restrictions, so the question of whether biofuels are environmentally acceptable has to be answered by stating that it depends on the economic arrangements prevailing in a given society.

The renewable energy systems explored in this book have been shown to be technically viable. Thus a natural follow-up question is whether they are also consistent with the organization of society prevailing in most countries of the current world. Up to a certain point, renewable energy can be—and often already is—introduced in the energy supply system within the current economic paradigm. Private investors and companies buy solar cells or shares in wind parks to show their greenness, and occasionally they also make money. However, the paradigm of the unrestrained market economy has the concept of economic growth as a foundational pillar, a concept that cannot continue to be compatible with basing energy supply systems on sources that—in one way or another (resource flow or sustainable utilization)—are finite. The same is of course true for any mineral resources and other growth parameters. But in order to discuss the eventual limits to growth, it is imperative to define precisely what is meant by growth. There may be immaterial growth than can continue forever, such as trade of fictitious shares that need not have any physical basis and may still allow trade and profits that shift hands indefinitely (e.g., the papers traded as future options).

The definition of growth used by economists is a straightforward invocation of GNP (gross national product) to measure the rate of growth. GNP is the sum of monetary value assigned to all those activities in a society that entail payment of money. For the present debate, one may disregard minor ingredients of discussion such as the way to include or omit subsidies, taxes, and payments to foreign workers, giving rise to GNP-related concepts such as the gross domestic product. Activities where money is not exchanged or transferred, such as much household work, are not included. The main problem is precisely that all monetary activities are included in the GNP, whether they are beneficial, neutral, or plainly detrimental. If I agree with my next-door neighbor to pay him a million dollars on even days, while he pays me a million dollars on odd days, then the GNP rises by a million dollars every day, as our handing money over the hedge is an economic transaction and thus has to be counted in the GNP. On this background, it is difficult to see why GNP *has* to rise for society to do well, but such is the paradigm that our elected politicians have chosen to borrow from the mainstream economists ruling the present world.

Before answering the question of whether growth is necessary, let me propose a measure of growth that would appear more reasonable. I call it the MDA or Measure of Desirable Activities. It is defined by the same monetary appraisal of activities as the GNP, but instead of just summing them up, each transaction cost is multiplied by a weight factor before performing the summation. The weight factor is positive if the activity is desirable for society as a whole (and not just those few who reap the revenue); it is zero if the activity is not contributing one way or the other (examples are advertising costs and the costs of financial transactions such as currency speculation through unnecessary money exchange transactions); and finally the factor is negative if the activity is detracting from the well-being of the society (e.g., by creating excess pollution or using more energy and resources than other ways of performing the same task or achieving the same end).

The idea of weighting the monetary value of activities in society was proposed by the Swedish military airplane constructor and political debater Olle Ljungström in the 1970s, where I first heard of it at conferences where he was speaking. At the time, there were many discussions of better ways to describe economic progress, and I am sure that other participants in the debate have had similar ideas. Today, most politicians worldwide seem to accept the GNP as a useful measure, although there is still occasional criticism at the grassroots level. Many voters believe the politicians and economist when they are told that without growth they would be without work and without money. But this has to be false: Unnecessary activities cannot be a basic condition for social well-being, as noted in the work of Mishan (1969), which criticized the conventional concept of "growth." If national economies were conducted according to the MDA, well-being would be the only success criterion, and if more well-being can be achieved with less work, then the rational action is to reduce working hours, as indeed happened during the part of the twentieth century when the welfare economies were formed in Scandinavia. Actually, working hours were at the time also reduced in countries where welfare was not high on the agenda.

There is consensus on the unlimited growth potential of non-material activities, but at present there are disputes over the limits to growth in activities associated with material consumption. The adherents of the presently favored economic growth paradigm naturally want to play down the finiteness of the Earth and its resources. The economic version of this is to claim that the market will always allow growth: If some material source is close to being exhausted, then substitution will take place, in pace with the relative prices that the markets assign to the traditional resource and its new substitute. The question is, of course, whether substitution is always possible. The growth proponents say, "Look at how metals have been replaced by plastic compounds," but they seem to forget that plastics are derived from fossil fuels, of which there is certainly a finite amount, as they were formed only during a very special period of time (some 10^8 years ago) in the Earth's history (see Sørensen [2012, chap. 7]). The view that substitution is always possible at a certain price is fundamentally wrong, but it seductively does appear to work on a short-term scale, and this is of course why the references to substitution have become prominent in current economic advice to politicians—substitution is not a concept that is supposed to be tested for durability.

One reason for insisting on economic growth is the trickle-down theory (or rather speculation) that if the total GNP grows, then even the poorest will experience some improvement. In practice, this may occasionally be true, but at the same time, disparity increases. Current societies believe that they can better deal with social unrest caused by the uneven distribution of wealth and well-being than nations could earlier in history, where excessive gaps between the rich and the poor were vehicles for revolutionary conflicts. Smoothing out differences in wealth were goals of the welfare economies established during the 1930s, notably in Scandinavia, to be achieved through taxation and redistribution of the money collected. Currently, these ideas of a "third way" between capitalism and socialism have lost momentum, and although voters in some countries are still uncomfortable by seeing beggars along their city streets, political advertising campaigns have gained widespread acceptance for the belief that a society where a few are rich and the rest are poor is a healthy state that just encourages the poor to work harder.

There are two concepts in the current economic paradigm that need scrutiny. One is the definition of "work," and the other is the fundamental question of what constitutes "social welfare." Work is a relatively new concept, replacing what could be called "meaningful activities." It was introduced in connection with the rise of capitalism in Scotland and surrounding regions (Anthony 1977). In earlier times, societies depended on contributions by its members, often involving toil, but still leaving room for a meaningful combination of leisure and needed activities. In regions of rich natural environments, such as the southern part of Africa, a mere 2–3 hours of daily activity per capita sufficed to provide the needed food and material requirements (see studies of the !Kung tribes by Lee [1979] and Wiessner [2002]). The free-market capitalism as well as its socialist counterpart defined the concept of "workers" as salaried human production factors, and the former used, according to Anthony, the Christian-Protestant claim that "idleness is a sin" to impose long hours of uninterrupted work as a condition for the working class of citizens. More parts of society became enrolled in the formal, monetary economy, and earning money became a basic condition of life—a carrot that people chased in their efforts to do as well as possible, feeling proud when their personal earnings were larger than those of the citizen next door. The alternative of maximizing welfare by an equitable distribution policy (as seen in the 1950s in Scandinavia, where the per capita earnings of the richest 10% of the population were only a factor of three above the earnings of the poorest 10%) has been rejected in most parts of the world. This has an impact on the debate about growth, because lack of redistribution of wealth makes salaried people and their spokespersons fear a halt to growth, which they believe will lead to widespread unemployment rather than to a generally relaxed work environment and reduced working hours (Illich 1978; Gemynthe 1998).

The other point worth discussion is what we mean by the term *welfare*. To the social democratic political parties of the 1930s (stressing that they were not communists and did not like dictatorships by Stalinists or by the Proletariat), *welfare* included a cluster of services accessible to all citizens—medical care, education, unemployment compensation, and social help to people involuntarily in need of help of one kind or another. Although many of these services are today replaced by insurance-type arrangements, there are still many who have not paid sufficient premium to be eligible for such insurance. In some cases, people unwisely used their money on nonbasic consumption, but in other cases, their jobs were so poorly paid that they could not afford insurance without sacrificing basic daily needs. Two tendencies of present economic behavior amplify these types of problems. One is the use of advertising that persuades people to buy things they do not need and often cannot afford. The other is the recurrent economic bubbles, often caused by banks lending out money to unnecessary consumption, money that the debtor cannot repay and for which there is no collateral once the bubble bursts, most often associated with overvalued real estate property suddenly losing value (and often compounded with high operational expenses for buildings due to the poor energy standards in most of the current building stock).

The question is whether proper welfare is just a return to the 1930 Scandinavian state of affairs. Probably not! There were traits of the Scandinavian situation then that we would not want to revive, and although some of the features of that time, such as solidarity between people, has been lost or replaced by competition and self-centered behavior, there are also positive changes, e.g., the furthering of equitable coexistence between sexes and races, that have become commonly accepted (at least in the so-called Western world) over the recent decades. More generally, there are many necessary ingredients of social welfare that go beyond securing individual people from falling through the bottom of the economic net, although this should of course be a priority in any case. These other items include tolerance across religious and political views, abolition of warfare as a solution to political problems, as well as cultivating the basic altruism said to lie buried in all of us.

Replacing work by meaningful activity may sound complex, but elements of the needed mindset already exists. Many salary earners are actually dedicated to their work and would perform it equally well without the demand to produce a profit to some shareholder. Examples include nonprofit organizations, but there are also creative people in all kinds of jobs, from teachers to technical consultants. The idea of funding industry and business by inviting investors to buy shares originated with the idea that the investors would choose a business to invest in from a personal belief that the product or activity would have an important future and benefit its customers. Current investments are often blurred as regards purpose, offering rapidly changing investment portfolios administered by some bank or similar institutions, with the investor often not knowing what would be produced, assumedly caring only about the stock value and dividends. Surely a society arranging its investment policy according to a quality assessment of the industries and businesses created—and in the new paradigm supported by public money (i.e., contributions from all members of society)—can do better than the banks and investment firms presently making the choices for us and creating the recurrent financial crises. It is not the purpose of this book to elaborate on new economic paradigms, but it needs to be stated that any comparisons of the energy systems considered in this book are highly dependent on the arrangements selected by the societies involved, and some of the proposed solutions make sense only in an atmosphere of collaboration between countries not seeking indiscriminate growth but, rather, showing consideration for the arguments made here regarding the need for an altered economic paradigm.

References

Anthony, P. 1977. *The ideology of work*. London: Tavistock.
Gemynthe, F. 1998. *Rigdom uden arbejde* (in Danish). Copenhagen: Chr. Ejler's Forlag.
Illich, I. 1978. *The right to useful unemployment*. Boston: Marion Boyars.

Lee, R. 1979. *The !Kung San: Men, women, and work in a foraging society.* Cambridge, UK: Cambridge University Press.

Mishan, E. 1969. *Growth: The price we pay.* London: Staples Press.

Petersen, A., C. Juhl, T. Pedersen, H. Ravn, C. Søndergren, P. Simonsen, K. Jørgensen, et al. 2001. Scenarier for samlet udnyttelse af brint som energibærer I Danmarks fremtidige energisystem. *Tekster fra IMFUFA Nr. 390* (in Danish). Roskilde University. Available at http://rudar.ruc.dk.

Sørensen, B. 2000. Editorial: The need for global modelling. *Int. J. Global Energy Issues* 13 (1–3): 1–3.

Sørensen, B. 2011. *Life-cycle analysis of energy systems: From methodology to applications.* Cambridge, UK: RCS Publishing.

Sørensen, B. 2012. *Fuel cells and hydrogen.* 2nd ed. Burlington, MA: Elsevier.

Sørensen, B., and P. Meibom. 2000. A global renewable energy system. *Int. J. Global Energy Issues* 13 (1–3): 196–276.

Wiessner, P. (2002). Hunting, healing, and *hxaro* exchange: A long-term perspective on !Kung (Ju/'hoansi) large-game hunting. *Evolution and Human Behavior* 23:407–36.

Appendix: Units and Conversion Factors

Unit Prefixes
Powers of 10

PREFIX	SYMBOL	VALUE
atto	a	10^{-18}
femto	f	10^{-15}
pico	p	10^{-12}
nano	n	10^{-9}
micro	μ	10^{-6}
milli	m	10^{-3}
kilo	k	10^{3}
mega	M	10^{6}
giga	G	10^{9}
tera	T	10^{12}
peta	P	10^{15}
exa	E	10^{18}

Units
SI Units

BASIC UNIT	NAME	SYMBOL	DEFINITION
length	meter	m	
mass	kilogram	kg	
time	second	s	

SI Units

BASIC UNIT	NAME	SYMBOL	DEFINITION
electric current	ampere	A	
temperature	Kelvin	K	
luminous intensity	candela	cd	
plane angle	radian	rad	
solid angle	steradian	sr	
amount[a]	mole	mol	

DERIVED UNIT	NAME	SYMBOL	DEFINITION
energy	joule	J	$kg \cdot m^2 \cdot s^{-2}$
power	watt	W	$J \cdot s^{-1}$
force	newton	N	$J \cdot m^{-1}$
electric charge	coulomb	C	$A \cdot s$
potential difference	volt	V	$J \cdot A^{-1} \cdot s^{-1}$
pressure	pascal	Pa	$N \cdot m^{-2}$
electric resistance	ohm	Ω	$V \cdot A^{-1}$
electric capacitance	farad	F	$A \cdot s \cdot V^{-1}$
magnetic flux	weber	Wb	$V \cdot s$
inductance	henry	H	$V \cdot s \cdot A^{-1}$
magnetic flux density	tesla	T	$V \cdot s \cdot m^{-2}$
luminous flux	lumen	lm	$cd \cdot sr$
illumination	lux	lx	$cd \cdot sr \cdot m^{-2}$
frequency	hertz	Hz	$cycle \cdot s^{-1}$

[a] The amount containing as many particles as there are atoms in 0.012 kg ^{12}C.

Conversion Factors

TYPE	NAME	SYMBOL	APPROXIMATE VALUE
energy	electron volt	eV	1.6021×10^{-19} J
energy	erg	erg	10^{-7} J (exact)
energy	calorie (thermochemical)	cal	4.184 J
energy	British thermal unit	Btu	1055.06 J
energy	Q	Q	10^{18} Btu (exact)
energy	quad	q	10^{15} Btu (exact)
energy	tons oil equivalent	toe	4.19×10^{10} J
energy	barrels oil equivalent	bbl	5.74×10^{9} J
energy	tons coal equivalent	tce	2.93×10^{10} J
energy	m^3 of natural gas		3.4×10^{7} J
energy	kg of methane		6.13×10^{7} J
energy	m^3 of biogas		2.3×10^{7} J

TYPE	NAME	SYMBOL	APPROXIMATE VALUE
energy	liter of gasoline		3.29×10^7 J
energy	kg of gasoline		4.38×10^7 J
energy	liter of diesel oil		3.59×10^7 J
energy	kg of diesel oil/gas oil		4.27×10^7 J
energy	m^3 of hydrogen at 1 atm		1.0×10^7 J
energy	kg of hydrogen		1.2×10^8 J
energy	kilowatt-hour	kWh	3.6×10^6 J
power	horsepower	hp	745.7 W
power	kWh per year	kWh/y	0.114 W
radioactivity	curie	Ci	3.7×10^8 s^{-1}
radioactivity	becquerel	Bq	1 s^{-1}
radiation dose	rad	rad	10^{-2} J·kg^{-1}
radiation dose	gray	Gy	1.0 J·kg^{-1}
dose equivalent	rem	rem	10^{-2} J·kg^{-1}
dose equivalent	sievert	Sv	1.0 J·kg^{-1}
temperature	degree Celsius	°C	$K - 273.15$
temperature	degree Fahrenheit	°F	$9/5$ °C+ 32
time	minute	min	60 s (exact)
time	hour	h	3600 s (exact)
time	year	y	8760 h
pressure	atmosphere	atm	1.013×10^5 Pa
pressure	bar	bar	10^5 Pa
pressure	pounds per square inch	psi	6890 Pa
mass	ton (metric)	t	10^3 kg
mass	pound	lb	0.4536 kg
mass	ounce	oz	0.02835 kg
length	angstrom	Å	10^{-10} m
length	inch	in.	0.0254 m
length	foot	ft	0.3048 m
length	mile (statute)	mi	1609 m
volume	liter	L	10^{-3} m^3
volume	gallon (US)	gal	3.785×10^{-3} m^3

Index

Printed and bound by CPI Group (UK) Ltd, Croydon, CR0 4YY

21/10/2024

01777089-0008